The Analysis of
Nutrients in Foods

FOOD SCIENCE AND TECHNOLOGY

A SERIES OF MONOGRAPHS

Maynard A. Amerine, Rose Marie Pangborn, and Edward B. Roessler, PRINCIPLES OF SENSORY EVALUATION OF FOOD. 1965.

C. R. Stumbo, THERMOBACTERIOLOGY IN FOOD PROCESSING, second edition. 1973.

Gerald Reed (ed.), ENZYMES IN FOOD PROCESSING, second edition. 1975.

S. M. Herschdoerfer, QUALITY CONTROL IN THE FOOD INDUSTRY. Volume I — 1967. Volume II — 1968. Volume III — 1972.

Hans Riemann, FOOD-BORNE INFECTIONS AND INTOXICATIONS. 1969.

Irvin E. Liener, TOXIC CONSTITUENTS OF PLANT FOODSTUFFS. 1969.

Martin Glicksman, GUM TECHNOLOGY IN THE FOOD INDUSTRY. 1970.

L. A. Goldblatt, AFLATOXIN. 1970.

Maynard A. Joslyn, METHODS IN FOOD ANALYSIS, second edition. 1970.

A. C. Hulme (ed.), THE BIOCHEMISTRY OF FRUITS AND THEIR PRODUCTS. Volume 1 — 1970. Volume 2 — 1971.

G. Ohloff and A. F. Thomas, GUSTATION AND OLFACTION. 1971.

George F. Stewart and Maynard A. Amerine, INTRODUCTION TO FOOD SCIENCE AND TECHNOLOGY. 1973.

Irvin E. Liener (ed.), TOXIC CONSTITUENTS OF ANIMAL FOODSTUFFS. 1974.

Aaron M. Altschul (ed.), NEW PROTEIN FOODS: Volume 1, TECHNOLOGY, PART A — 1974. Volume 2, TECHNOLOGY, PART B — 1976. Volume 3, ANIMAL PROTEIN SUPPLIES, PART A — 1978.

S. A. Goldblith, L. Rey, and W. W. Rothmayr, FREEZE DRYING AND ADVANCED FOOD TECHNOLOGY. 1975.

R. B. Duckworth (ed.), WATER RELATIONS OF FOOD. 1975.

A. G. Ward and A. Courts (eds.), THE SCIENCE AND TECHNOLOGY OF GELATIN. 1976.

John A. Troller and J. H. B. Christian, WATER ACTIVITY AND FOOD. 1978.

A. E. Bender, FOOD PROCESSING AND NUTRITION. 1978.

D. R. Osborne and P. Voogt, THE ANALYSIS OF NUTRIENTS IN FOODS. 1978.

In preparation
John Vaughan (ed): Food Microscopy

The Analysis of Nutrients in Foods

by

D. R. OSBORNE

**Unilever Research
Bedford, England.**

and

P. VOOGT

**Unilever Research
Vlaardingen/Duiven
Zevenaar, The Netherlands**

1978

ACADEMIC PRESS

London   New York   San Francisco

A Subsidiary of Harcourt Brace Jovanovich, Publishers

ACADEMIC PRESS INC. (LONDON) LTD.
24/28 Oval Road
London NW1

United States Edition published by
ACADEMIC PRESS INC.
111 Fifth Avenue
New York, New York 10003

Library of Congress Catalog Card Number: 77-75365
ISBN: 0-12-529150-7

Printed in Great Britain by
Willmer Brothers Limited, Birkenhead

Preface

Having been hunter, fisher and farmer, living off a restricted area of surrounding countryside for thousands of years, man has over the past one hundred years quite suddenly been exposed to the industrial revolution and a population explosion, which have produced dramatic changes in both where and how he lives.

For example, the shift in population from rural to urban life in industrialised countries has been such that it is no longer possible to satisfy all the demands of society without recourse to the extensive use of modern agricultural practice and food technology. Without this it would not be possible to provide fresh or processed foods of good quality on a large enough scale, to preserve foods against deterioration, to prepare foods for the convenience of the consumer and, most important in the context of this book, to provide a general diet of satisfactory nutritive value.

Most of us tend to take for granted the wide variety and volume of food available to us today. We should recognise, however, that advances in agriculture and food technology with its techniques of processing, storage and distribution, have been as important to us in maintaining our nutritional standards during a period of rapid changes in society, as advances in sanitation, medical practice and pharmacology have been in eradicating disease.

Whilst recognising that continuing changes in agricultural practice and food technology will occur to improve the supply of raw materials, increase the variety of products and facilitate storage and domestic use of foods, we should not become complacent about the effects that they and our eating habits may be having on our nutritional wellbeing.

74098

One of the key elements in monitoring our general nutritional status is reliable information on the nutrient composition of foods and it is in this area in particular that the food analyst has a major contribution to make to the advancement of nutritional research. The primary purpose of this book is therefore to provide the analyst with a reliable and generally applicable set of analytical procedures for the determination of those nutrients for which the significance in the human diet is known.

These methods are set out in standard form in Part II of this book. Because the range of food products now available is enormous and the resources of the analyst often limited, attention has been given to replacing or providing alternatives to some of the more tedious of the classical biological and chemical procedures. For example, new high performance chromatographic methods are included for certain vitamins and atomic absorption methods replace the classical chemical procedures for metals. These techniques will allow the analyst who has access to modern instrumentation to deal with large numbers of samples in a limited time. Thus the book is intended to reflect modern developments, but at the same time it recognises the widely differing operational conditions of individual analysts and hence the continuing need for and value of less sophisticated but well tested methods of analysis.

Reliable and efficient analytical procedures are not enough, however, if the analyst is to provide an effective service and contribute fully to nutrition research and development. He must, wherever possible, increase his general understanding of nutrition so that he can appreciate the objectives of any nutritional work before doing his analyses, as it is only with such information that he will be able to judge which are the most important analyses to do, put problems into their correct perspective, apply the appropriate tests and attempt to interpret data intelligently. Failure to do this will invariably result in a waste of valuable technical resources and poorly defined answers to the problem at hand.

For example, a working knowledge of data tables on the levels of nutrients in foods and average food consumption, coupled with information on the stability of various nutrients under conditions normal in agricultural practice, food processing, storage and household cooking, should form an essential part of the background against which the analyst can judge how important it is to carry out certain analyses.

Take, for example, vitamin C in milk and potatoes. Data tables for raw and processed milk will show that there is often considerable destruction of vitamin C during processing, but they show also that the total amount of the vitamin initially present in raw milk is small and that milk is, therefore, under normal patterns of food consumption, unlikely to be regarded as a significant dietary source of vitamin C. Therefore, processing losses of vitamin C from milk are nutritionally less important and may not require frequent monitoring. By contrast, potatoes supply one third to one half of the average vitamin C intake of Europeans in winter. Losses of vitamin C through processing potatoes could therefore be serious and more frequent monitoring may be justified.

In order to help the analyst to make a full and effective contribution to nutrition research and development, Part I of the book contains, in addition to further analytical background, some elementary information on the chemistry, biochemistry and biological role of the main macro- and micro-nutrients and introductory material on food composition, intake of nutrients and the interpretation of nutritional data. This material is in no way comprehensive but tries to illustrate the need for and value of the analyst becoming more familiar with the general background to nutritional issues before undertaking the practical work involved. This process can begin easily with this book but the reader is also strongly advised to read further into the subject via appropriate books and other literature, such as those provided in the classified bibliography at the end of Part I.

In summary, the book should be looked upon as serving two functions, the primary one being to provide the analyst with the tools of his trade in the form of a comprehensive set of relevant methods and a secondary but no less important one of introducing the analyst to the broader aspects of nutrition in a way that will help and encourage him to become an active and effective partner rather than servant of nutritional research and development.

Finally, the authors would like to acknowledge gratefully the efforts and co-operation of their many co-workers at the Unilever Research Laboratories at Colworth and Duiven, without which this book could not have been produced.

D.R.O.
P.V.

April, 1978

Contents

Preface v

PART I   THE CHEMISTRY, BIOLOGICAL ROLE AND
             ANALYSIS OF NUTRIENTS IN FOOD

CHAPTER 1   Chemistry and Biological Role of Macronutrients . 5
CHAPTER 2   Chemistry and Biological Role of Micronutrients . 22
CHAPTER 3   Analysis of Nutrients in Food 43
CHAPTER 4   Recommended Intake of Nutrients and Interpretation
             of Nutritional Data 56
CHAPTER 5   Food Composition Tables 79
APPENDIX   Reference Works for further reading . . . 91

PART II   METHODS FOR THE ANALYSIS OF NUTRIENTS
              IN FOOD

CHAPTER 6
Introduction 103
Section 1   General Sample Preparation 105
Section 2   Moisture and Total Solids 107
Section 3   Proteins and Nitrogenous Compounds . . . 113
Section 4   Carbohydrates 130
Section 5   Lipids 155
Section 6   Ash, Elements and Inorganic Constituents . . 166
Section 7   Fat-Soluble Vitamins 183
Section 8   Water-Soluble Vitamins 201
Section 9   Calculation of Calorific Value 239

Index 241

PART I

CHAPTERS 1–5

The Chemistry, Biological Role, and Analysis of Nutrients in Food

CONTENTS

Chapter 1   Chemistry and Biological Role of Macronutrients

1.1 Proteins 5
1.2 Carbohydrates 9
1.3 Lipids 12
1.4 Calorific value 18
1.5 Water 20

Chapter 2   Chemistry and Biological Role of Micronutrients

2.1 Minerals 22
2.2 Vitamins 28

Chapter 3   Analysis of Nutrients in Food

3.1 Introduction 43
3.2 Sampling—Errors and Procedures 44
3.3 Analysis for Macronutrients 46
3.4 Analysis for Micronutrients 51

Chapter 4   Recommended Intake of Nutrients and Interpretation
of Nutritional Data

4.1 Recommended Intake of Nutrients 56
4.2 Reasons for the Differences in Recommended Daily Intake 67
4.3 Interpretation of Nutritional Data 74

Chapter 5   Food Composition Tables

5.1 The Value and Uses of Food Composition Tables . . 79
5.2 Sources and Scope of Data 80
5.3 Conversion Factors, Abbreviations, and Tables . . . 81

Appendix   Reference Works for further reading   91

CHAPTER 1

Chemistry and Biological Role of Macronutrients

1.1 Proteins

Proteins are polymers of which the basic units are amino or imino acids. In a protein molecule hundreds, sometimes thousands, of amino or imino acids are joined together by a characteristic linkage, the peptide bond. The basic protein structure is represented in Fig. 1 in which R can vary depending on the amino acid units from which the chain is formed, and the sequence of amino acid groups is developed in a specific way for a given protein.

In addition to the specific sequence of primary linkages which forms the backbone of protein molecules, other secondary physical or chemical linkages are present; for example, the linking of two or more chains and conformational factors make possible a multiplicity of structural differences between proteins. Thus the physical, chemical, and nutritional properties of different proteins vary with their quantitative amino acid composition, the sequence and cross-linking of the amino acid units in the protein molecule, and conformational arrangement.

Proteins in the human diet come from both animal and vegetable sources, the most important being meat, fish, milk, eggs, cereals, legumes, seeds, and

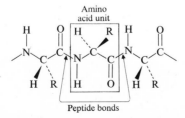

FIG. 1. The basic structure of proteins.

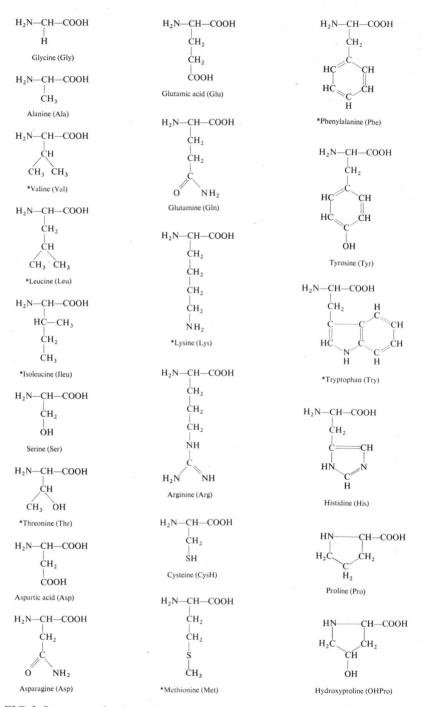

FIG. 2. Structures of amino acids and imino acids; the names of the 'essential' amino acids are marked with an asterisk(*).

nuts. When these are eaten, the proteins are digested by hydrolytic enzymes of the gastrointestinal tract and are absorbed into the bloodstream as amino acids. These amino acids are used in the synthesis of new proteins needed for growth, maintenance, and repair of body cells. Some of the amino acids required can be made in the body as the need arises but others can be obtained only from food. These latter are called essential amino acids. However, the non-essential amino acids are not necessarily less important biologically.

Twenty amino acids commonly occur in food, and eight of these are essential. These are shown diagrammatically in Fig. 2. With two exceptions these substances have a primary amino function ($-NH_2$) and a carboxyl function ($-COOH$) joined to the same carbon atom; hence they are termed α-amino acids. The two exceptions, proline and hydroxyproline, are α-imino acids (see Fig. 3). Egg and human milk proteins provide all the essential amino acids for normal growth and healthy life processes, provided that they are taken in adequate amounts. On the scale of nutritive value, they are at the top.

These proteins contain the amino-acids in the correct proportion for the body's need. If such a protein is fed under experimental conditions to growing animals, virtually 100% of the protein absorbed is used for protein synthesis, and none is diverted to energy production; hence the protein is said to have a Biological Value (BV) of 100. No other protein is quite as good. Cow's milk, fish, and meat come next, followed by the plant proteins, i.e. wheat, rice, beans, and nuts. For example, cow's milk protein is short of methionine and cystine and has a BV of 75; wheat protein is deficient in lysine and has a BV of only 50. In the latter case lysine is termed the 'limiting amino acid'.

Human diets do not of course consist of individual proteins, but are made up of mixtures of very many proteins. This mixing has important implications for us, since it will often happen that the mixing of different proteins of lower BV will produce a mixture with a higher BV than we might expect from the average of the separate BV's. If this occurs the proteins are said to complement one another nutritionally. The principle of complementation is best illustrated diagrammatically (Fig. 4). A practical example is bread having a BV of 50 and cheese having a BV of 75, yet a mixture of $3\frac{1}{2}$ parts bread to 1 part of cheese has a BV of 75. The reason is that the shortage of lysine in bread is made up by the surplus of lysine in cheese.

The diagram (Fig. 4) will also help to explain why complementation does not occur if both proteins in a mixture are deficient in the same amino-acid.

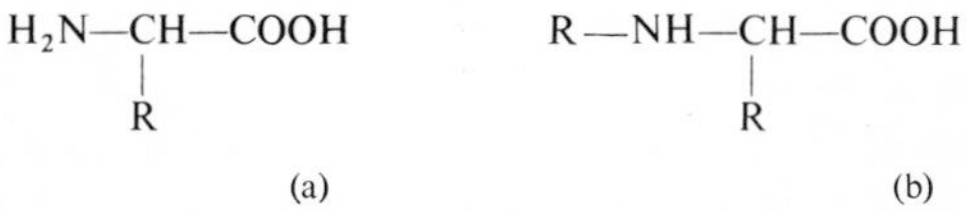

$$H_2N-CH-COOH \qquad\qquad R-NH-CH-COOH$$
$$\qquad\ \ |\qquad\qquad\qquad\qquad\qquad\qquad |$$
$$\qquad\ \ R\qquad\qquad\qquad\qquad\qquad\qquad R$$

(a)   (b)

FIG. 3. General formulae of (a) α-amino acids and (b) α-imino acids.

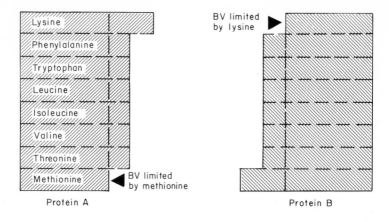

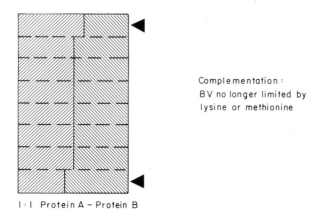

FIG. 4. Principle of complementation of proteins.

Thus a mixture of maize (BV 36) and bread (BV 50), both limited by lysine, has a BV of 43, the average of the individual values. Also, complementation between proteins does not occur if the proteins are eaten at different times. To ensure complementation, the two proteins must be eaten at the same time at the same meal, because the body has little capacity to store unused amino acids but uses them as sources of energy.

We have so far defined protein quality in terms of Biological Value, but readers will almost certainly encounter other indices of protein quality. Biological Value is defined strictly as 'the percentage of absorbed nitrogen retained in the body of a growing animal for growth and maintenance', where 'nitrogen' is taken as an easily measured index of protein content. Not all proteins are equally absorbed; in other words, some proteins are more readily digestible than others. These differences are taken into account in another index of quality, Net Protein Utilization (NPU), which is the percentage of *ingested* nitrogen utilised for growth and maintenance. A third popular index is Protein Efficiency Ratio (PER), which is the gain in body weight of a group of test animals divided by the weight of protein consumed. The quality of soy protein, beef protein, and milk protein expressed in these three ways is as follows:

	BV	NPU	PER
Soy	75	69	2.3
Beef	76	76	3.2
Milk	85	83	2.9

1.2 Carbohydrates

The carbohydrates of human diet are primarily derived from plant materials, e.g. cereals and products derived from them such as flours, vegetables, sugar, and preserves. The most important exception to this in nutritional terms is the animal-derived milk sugar lactose. Carbohydrates supply a major portion of man's energy. Current nutritional opinion tends to favour a level of carbohydrate in diet which supplies 50–65% of the total energy requirement. Nevertheless, ecologic and economic factors produce wide variations in the percentage of total energy from carbohydrate in diet (i.e. in extreme cases 20–80%), which are still consistent with normal health.

Food carbohydrates occur in various forms (Fig.5). Firstly, they are found in relatively small amounts as monosaccharides (e.g. glucose, fructose, galactose, ribose). Secondly, they either occur naturally or are added to food in substantial amounts as disaccharides (e.g. mainly sucrose and lactose, with smaller amounts of others such as maltose). Thirdly, they may be present in small amounts as dextrins formed from the partial breakdown of poly-saccharides. Fourthly, they are present in substantial amounts as polysaccharides (e.g. starch or its animal equivalent glycogen, cellulose, hemicelluloses, pectins, and others).

All these carbohydrate forms, however, are not equally 'available' in the sense that they are metabolised and make the same contribution to man's energy supply. For example, the polysaccharide starch, which is a major contributor to food carbohydrates, is considered to be 'available', in that it is broken down by enzymes in the mouth and small intestine which cleave α-(1-4)

glycosidic linkages and is degraded primarily by repeatedly splitting off
disaccharide (maltose) units. These, together with other disaccharides already
in the diet (e.g. sucrose, lactose) are absorbed into the mucosa of the small
intestine where enzymes split them into monosaccharides (glucose, fructose,
galactose). They are then absorbed into the bloodstream and transported to
body tissues for synthesis or energy production.

Conversely, the polysaccharide cellulose, which serves as the main structural
component of plant tissues, is considered to be 'unavailable'. The human

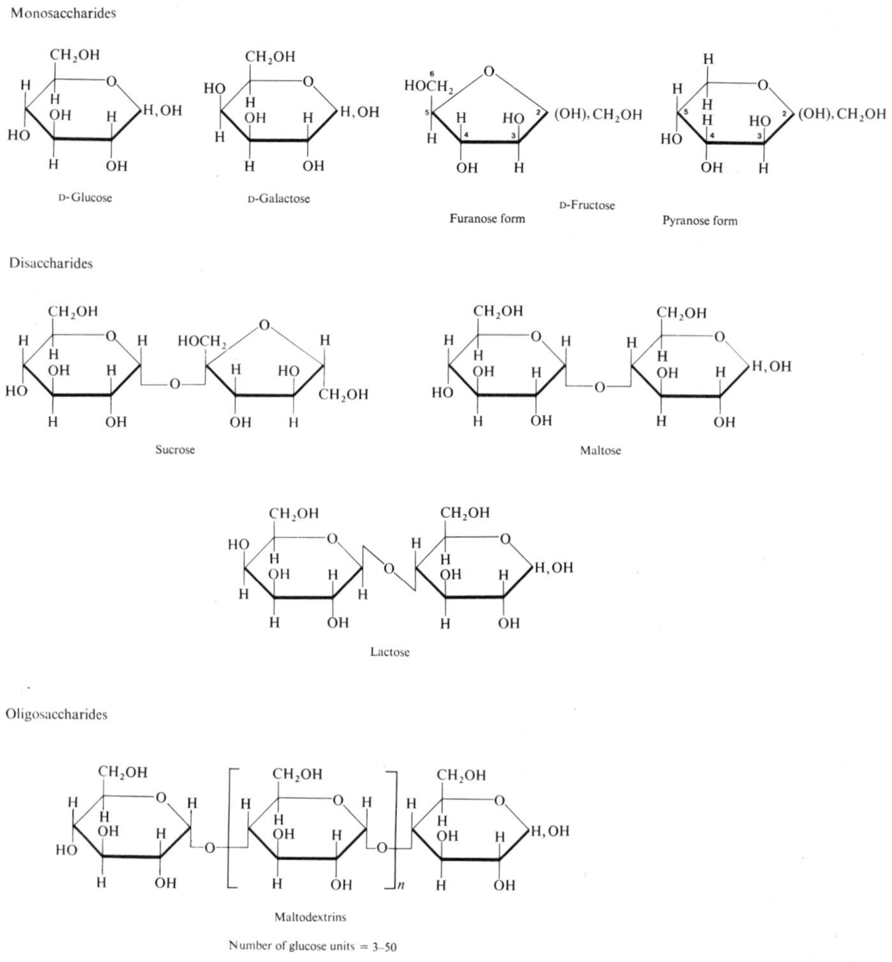

FIG. 5. Structures of some important food carbohydrates.

Polysaccharides

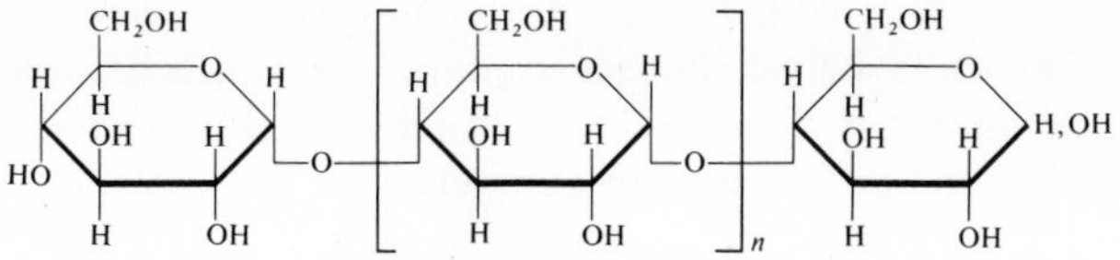

Starch unbranched amylose α-(1–4) glycosidic linkages

Approximate number of glucose units = 2000

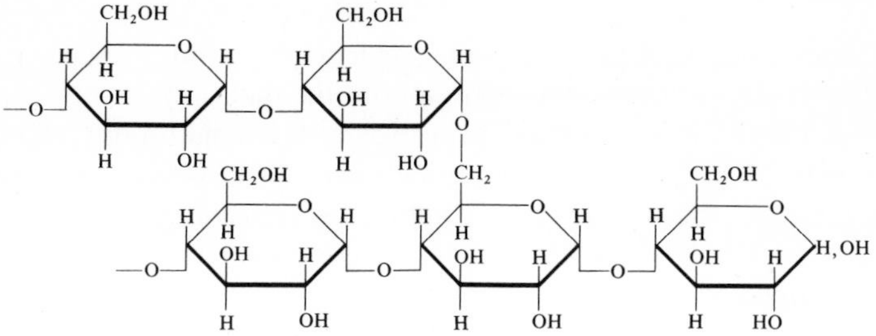

Starch branched amylopectin α-(1–4) and α-(1–6)
glycosidic linkages
Approximate number of glucose units = 100 000

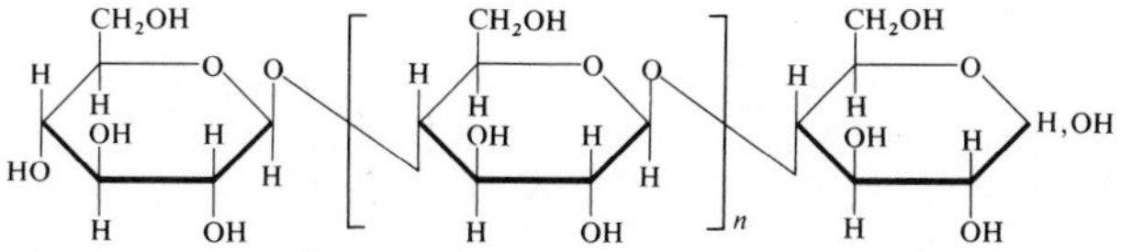

Cellulose β-(1–4) glycosidic linkages
Approximate number of glucose units = 6000

FIG. 5. (*continued*)

digestive tract does not secrete any enzyme capable of splitting β-(1-4) glycosidic linkages and most of it along with other 'unavailable' carbohydrates (hemicelluloses etc.) and indigestible matter such as lignin (collectively termed 'dietary fibre') passes through the body unchanged providing roughage and bulk essential for the proper functioning of the lower intestinal tract. Though intestinal bacteria may here break down and utilise variable amounts of this unavailable carbohydrate, this is not considered to contribute significantly to man's energy supply.

Although there is still some debate as to whether some available/non-available carbohydrate may be more available/non-available than others, and variations have been observed between different species in their ability to use unavailable carbohydrate, the following classification of food carbohydrates is generally accepted. The 'available' carbohydrate fraction in food is comprised of the sum of the monosaccharides, disaccharides, dextrins, starches, and glycogens. The 'unavailable' carbohydrate is comprised of the sum of naturally occurring cellulose, hemicelluloses, gums, and pectins and other polysaccharides that may be added to foods in low levels as technological aids (e.g. agars, carageenans, locust bean gum, etc.).

1.3 Lipids

Meats, eggs, dairy products, and fats, particularly in butter, margarine, and cooking oils, are primary sources of lipid in the diet. Lipids supply a major portion of man's energy supplies, giving weight-for-weight more than twice as much energy as proteins or carbohydrates. Just as the proportion of carbohydrate in the human diet is influenced by ecological and economic factors, so is the level of lipid which varies considerably from 6–10% in underdeveloped and overpopulated areas to 35–45% in the more prosperous countries. Current nutritional opinion tends to favour a level of lipid in diet equal to 25–35% of the total energy requirement.

Lipids are a heterogeneous group of naturally occurring substances which are insoluble in water but soluble in a range of organic solvents (e.g. chloroform, ether, hexane, etc.). There are three main classes of lipids: simple lipids, compound lipids, and derived lipids. Their classification, derivation, structure, and occurrence are shown in Tables 1–3.

To summarise, although lipids are a heterogeneous and complex group of naturally occurring substances, all of which have important physiological roles to play, in terms of the total dietary intake of lipid by far the largest components (i.e. usually $>95\%$) are the glycerides, with smaller amounts ($<10\%$) of waxes, phospholipids, sterols, and fatty acids, and only trace amounts of other lipids.

When ingested, fat (mainly as triglyceride) passes through the stomach and enters the duodenum where it is hydrolysed by enzymes (e.g. lipases) to produce fatty acids, monoglycerides and glycerol. These together with bile salts produce conditions in which fat can be emulsified into very small droplets which further aids digestion and absorption into the body from the mucous membrane of the small intestine. Here further digestions of the fat may occur and the process of re-synthesis of new triglycerides from fatty acids in the bloodstream of the body takes place. Thus in the process of digestion and absorption the glycerides of the food lose some of their identity and are replaced by triglycerides more characteristic of the species ingesting them.

After absorption, fats are then transported as lipoprotein complexes in the blood, either directly, or indirectly via the liver where they undergo further

TABLE 1. Classification, structure, and occurrence of simple lipids

Classification	Derivation	General structure	Substituents (R)
Glycerides (oils or fats)	Esters of monocarboxylic acids (the fatty acids) and glycerol	CH_2OCOR CH_2OH | | $CHOH$ $CHOCOR$ | | CH_2OH CH_2OH (Monoglycerides)	
		CH_2OCOR CH_2OCOR | | $CHOCOR$ $CHOH$ | | CH_2OH CH_2OCOR (Diglycerides)	C_nH_{2n+1} C_nH_{2n-1}
		CH_2OCOR | $CHOCOR$ | CH_2OCOR (Triglycerides)	C_nH_{2n-3} C_nH_{2n-5} C_nH_{2n-7} (For typical structures of **R** and **OR′** see derived lipids, Table 3)
Waxes	Esters of long-chain and alicyclic monohydroxy alcohols and fatty acids	$RCO(OR′)$	**(OR′)** = cholesteryl and others

General note on occurrence. Mixed triglycerides are the major components of natural oils and fats. Mono- and di-glycerides are present in natural oils and fats only in trace amounts. Processed fats may contain up to 20% of mono- and di-glycerides for technological purposes, e.g. their emulsifying properties. Waxes are widely distributed in nature but as a rule are not abundant in that they are usually present both in animals and in plants only as a surface layer.

TABLE 2. Classification, derivation, structure, and occurrence of compound lipids

Classification	Derivation	General structure	Substituents (R)
Phospholipids (i) Phosphoglycerides	Diglycerides containing phosphoric acid attached to (a) a nitrogenous base, most commonly choline, ethanolamine, and serine, or (b) polyhydric alcohols	CH_2OCOR^1 $\|$ R^2COOCH $\|\quad\quad O$ $\quad\quad\ \|\|$ $CH_2O-P-OR^3$ $\quad\quad\quad\|$ $\quad\quad\quad O_-$	$R^1 = $ Saturated fatty acid $R^2 = $ Unsaturated fatty acid $R^3 = -CH_2CH_2\overset{+}{N}(CH_3)_3$ (Lecithins) $= -CH_2CH_2\overset{+}{N}H_3$ $\quad\quad\quad\quad\quad\quad\quad$ (Cephalins) $= -CH_2\overset{+}{C}HNH_3$ $\quad\quad\quad\ \|$ $\quad\quad\ COOH$ (Inositol phosphatides) $= -CH_2-CH(OH)-CH_2OH$ (Glycerol phosphatides)
	Also related ethers	$CH_2OCH{=}CHR^1$ $\|$ R^2COOCH $\|\quad\quad O$ $\quad\quad\ \|\|$ $CH_2O-P-OR^3$ $\quad\quad\quad\|$ $\quad\quad\quad O_-$	$R^3 = -CH_2CH_2-\overset{+}{N}(CH_3)_3$ (Plasmalogens)

TABLE 2. (continued)

Classification	Derivation	General structure	Substituents (R)
(ii) Sphingolipids	Phosphatidylcholine derivatives of the long-chain unsaturated amino-alcohol sphingosine	$HO-CH-CH=CH[CH_2]_{12}CH_3$ $H-C-NH-CO[CH_2]_{22}CH_3$ $CH_2O-\overset{O}{\underset{O_-}{\overset{\|}{P}}}-OR^3$	$R^3 = -CH_2CH_2\overset{+}{N}(CH_3)_3$ (Sphingomyelins)
Glycolipids	Sphingolipids which contain a carbohydrate usually galactose in place of phosphatidylcholine	$HO-CH-CH=CH[CH_2]_{12}CH_3$ $H-C-NH-CO[CH_2]_{22}CH_3$ CH_2	Cerebroside or β-Galactolipid

General note on occurrence. Phospholipids usually make up 1–2% of many natural vegetable oils and higher percentages of natural animal fats, being found in nerve, brain, liver, kidney, and heart tissues. Another important source is egg yolk which contains 20% phospholipid which accounts for many of its special culinary properties. Only trace amounts of derived lipid are found in processed fats because they are removed during refining. Low levels of lecithins are commonly added to food systems as emulsifiers.

TABLE 3. Classification, structure, and occurrence of derived lipids

Classification	General Structure	Specific Examples	Names
Fatty acids (i) Saturated*	$C_nH_{2n+1}COOH$		Palmitic acid
			Stearic acid
(ii) Unsaturated	$C_nH_{2n-1}COOH$		Oleic acid
(iii) Polyunsaturated†	$C_nH_{2n-3}COOH$		Linoleic acid
	$C_nH_{2n-5}COOH$		Linolenic acid
	$C_nH_{2n-7}COOH$		Arachidonic acid
(iv) Others e.g. Cyclic fatty acids			Sterculic acid
Unsaturated monohydroxy fatty acids			Ricinoleic acid

TABLE 3. (*continued*)

Classification	General Structure	Specific Examples	Names
Alcohols (i) Glycerol	$\begin{array}{l}CH_2OH\\ \mid\\ CHOH\\ \mid\\ CH_2OH\end{array}$		
(ii) Long-chain alcohols	$C_nH_{2n+1}OH$ e.g. $C_{16}H_{33}OH$		Cetyl alcohol
	$C_{30}H_{61}OH$		Myricyl alcohol
(iii) Sterols			Cholesterol
			Ergosterol

Other fat-soluble materials (e.g. Fat-soluble vitamins A, D, E, and K; for structures see section on vitamins)

General note of occurrence. Free fatty acids are present in foods only in trace amounts unless hydrolytic or oxidative deterioration of the food has taken place, in which case the food is likely to be unacceptable owing to rancidity. Sterols occur in the fats of plants and of animals but usually at levels less than 0.5%. Other alcohols and fat-soluble vitamins are present only in trace amounts.

*Principally palmitic and stearic but many others in smaller amounts. Usually unbranched and containing an even number of carbon atoms mainly from C_4 to C_{20} but up to C_{30}.

†As indicated, but others ranging from C_{10} to C_{24} with one to six double bonds.

breakdown and re-synthesis, to storage depots in the adipose tissue where they serve as a reservoir of energy or synthetic material.

Complete exclusion of fat from an otherwise adequate diet induces in animals cessation of growth, scaliness of skin, impaired reproduction, and kidney damage. These abnormalities are cured or prevented by feeding appropriate amounts of the unsaturated fatty acids linoleic acid and arachidonic acid. It appears that, although animal tissues are capable of synthesising most saturated and unsaturated fatty acids, they cannot produce the unsaturated fatty acid series having a double bond in the 6-position unless a dietary precursor is furnished. Arachidonic acid alone or in conjunction with linoleic acid are therefore considered essential (hence the term essential fatty acid) for normal development, particularly in the young animal where body reserves of fat are small.

Fat is thus a necessary component of living tissues and essential in human nutrition. Because it can be stored and can be mobilised, it is the prime reserve material for the body. A balanced intake is also essential to ensure the dietary supply of essential fatty acids and the fat-soluble vitamins A, D, and E.

1.4 Calorific Value

Energy consumption by the body is related both to heat energy associated with basic body function and temperature and to mechanical energy associated with the movement of organs and limbs. The energy value of food is measured in terms of the kilocalorie (kcal) which is a physical unit of heat. The mechanical heat equivalent is the kilojoule (1 kcal = 4.18 kilojoules).

Even when at rest the human body needs energy. The amount required has been determined experimentally to be about 1 kcal per kilogram of body weight per hour, or 1500–2000 kcal per day. However, this varies with an individual's metabolism and may be as low as 1200 kcal or as high as 2200 kcal. Thus a large part of human energy consumption via food is used just for maintaining essential life processes and body temperature.

Over and above satisfying the basic energy requirements of the resting body we need extra energy to be able to carry out the movement associated with our daily jobs and leisure activities. For example, a secretary engaged in light office work would require an extra 30 kcal per hour, a plumber engaged in light manual work would require 150 kcal per hour, and a lumberjack engaged in heavy manual labour would require an extra 400 kcal per hour.

When the body derives energy from food it is less than the amount of energy produced when the food is burned (completely oxidised) in a calorimeter. This is because the calorie-producing nutrients, which are mainly protein, fat, and carbohydrate, are not completely digested, absorbed, or oxidised to yield energy in the body. For example, the part of the protein in diet that is not used for the formation of new proteins may be used as an energy source, and the end-products of this protein metabolism are urea, uric acid, creatinine, and some amino-acids which are excreted from the body.

Thus, in a calorimeter, protein is completely converted into carbon dioxide, ammonia, oxides of nitrogen and sulphur, and water, and yields 5.6 kcal per gram, but in the body the amount of energy available is only 4.0 kcal per gram. Similarly, the calorific values for fat and starch, as measured by physical techniques, are 9.4 and 4.2 kcal per gram respectively, but the physiological energy values are 9.0 and 4.0 kcal per gram respectively. Allowances are made for this when calculating the calorific value of food.

Since no two foods and no two people are exactly the same, the physiological correction factors are based on averages and do not have the same accuracy as the values for protein, fat, and carbohydrate found by chemical means. This has resulted in many sets of conversion factors being derived by different workers in the field through using different experimental approaches to the problem.

Another complicating factor is the 'availability' of calories in certain food ingredients; i.e. there is often a difference between the number of calories that a diet would provide where the protein, fat, and carbohydrate in it completely digested, and the number of calories that it does in fact provide. This is mainly due to the so-called 'unavailable carbohydrates' which are naturally contained in plant food.

The calculation of calorific value is thus not straightforward and there is no clear-cut answer to it. It is, however, important to have a consistent approach to the interpretation of data, and for the analyst to state the basis of his calculations. In the absence of agreed international rules, the values in Table 4 are used in this manual.

TABLE 4. Energy values for gross nutrients.

	Energy (kcal/g)
Protein	4.0
Fat	9.0
Carbohydrate	3.75*

*Available carbohydrate expressed as mono-saccharide. This includes mono- and di-saccharides, dextrins, starches, and glycogens, but excludes crude fibre, indigestible carbohydrate and structuring agents, modified celluloses, alginate, pectin, etc., which should be considered as unavailable. When the chemical composition of carbohydrate fraction is not known with certainty, values quoted should be noted as approximate.

1.5 Water

Even a brief introduction to the topic of nutrition would be incomplete without some reference to water. Water is not a nutrient in the sense that it functions as a fuel, but it is a nutrient in the sense that it accounts for about half of total body-weight and three-quarters of lean body-weight, and without it the body cannot function and survive.

The amount of water that the body requires depends amongst other things upon its environment, its condition, and the work that one does. Generally the daily requirement is about 2.5 litres and originates from the intake of drinks (1.25 litres), from water associated with food (0.9 litre), and water that arises from the digestion or oxidation of food in the body (0.35 litre); i.e. every gram of carbohydrate, protein, and fat releases about 0.6, 0.4, and 1.1 gram of water respectively during body oxidation. Thus a normal diet providing 2700 kcal (11 297 kilojoules), consisting of 420 g of carbohydrate, 70 g of protein, and 80 g of fat, delivers by oxidation about 0.35 litre of water.

For the healthy adult the intake of water is balanced by the loss of water via urine, faeces, and evaporation from the lungs and skin as summarised in Table 5.

Water has unique properties; without it life as we know it on this planet could not exist. For example, its high dipole moment makes it a good solvent for polar solutes such as amino-acids and essential electrolytes, allowing for their easy transport through the body. Its specific heat is such that it acts as an effective heat sink in the maintenance of body temperature at 37°C. Its latent heat of evaporation is such that cooling of the body can be achieved by evaporation from lungs and skin.

As stated previously, water accounts for about half of total body weight, and it is present for several functions. About 60% of it is intracellular water and functions as important building material for the cells. A further 30% of it is intercellular water and its function is to transport nutrients and metabolites from the blood vessels to the cells and vice versa. The smallest part, approximately 10%, is found in the vessel system (blood, lymph) of the body

TABLE 5. Body water balance

(%)	Loss of water (litres)		Total	Input of water (litres)		(%)
20	Evaporation from lungs	0.5		0.35	Oxidation product of food	
16	Evaporation from skin	0.4	2.5		digestion	14
60	Losses through urine	1.5	litres	0.90	Water associated with food	36
4	and faeces	0.1		1.25	From drinks	50

and is called intravessel water. This is the main circulatory water and its function is to transport oxygen, nutrients, and metabolites to the appropriate parts of the body (kidneys, lungs, etc.) and to regulate body temperature. Though loss of water via urine is about 1.5 litres per day, the internal circulation of water through the body is much greater. For example, the amount of water that passes through the kidney system each day (150 litres) is about two times man's body weight; only 1.5 litres is separated out as urine.

Overall, water is probably the most variable and versatile 'nutrient' in the body. For example, it can be lost in large amounts during several hours of hard work in the full sun, but the body adjusts temporarily and functions effectively by drawing on buffer stocks of intercellular water, which can be replaced in due course by an appropriate drink.

CHAPTER 2

Chemistry and Biological Role of Micronutrients

2.1 Minerals

All forms of living matter require many minerals for their life processes. Virtually all the elements of the Periodic Table have been found in living cells, although not all are necessarily essential to life. The study of mineral nutrition is complex, and although it is convenient to discuss each element individually, it is important to recognise that, just as proteins, carbohydrates, and fats do not play independent roles in human nutrition, the minerals or inorganic nutrients are inter-related and balanced against one another. For example, calcium and phosphorus are in a defined relationship in the formation of bones and teeth. Sodium, potassium, magnesium, phosphate, and chloride ions serve individual and collective purposes in the control of body fluids. Many elements act alone or in conjunction with others as catalysts for essential enzymic processes in the body.

The animal body requires seven minerals in relatively large (gram) amounts (i.e. calcium, sodium, magnesium, potassium, phosphorus, chlorine, and sulphur), and at least seven in trace amounts (cobalt, copper, iodine, iron, manganese, molybdenum, and zinc). In addition, although not fully established as essential, fluorine appears to have an important prophylactic action in bones and teeth, chromium has been claimed to be a glucose tolerance factor, and selenium has been shown to protect against liver necrosis and other disorders in animals. There are also more recent reports that vanadium, tin, nickel, silicon, and several others are important in the proper development of experimental animals such as the rat but the importance of these in human nutrition terms has yet to be fully established.

For the purpose of giving a simple and brief introduction to the topic of minerals and human nutrition, the minerals can conveniently be discussed under three main headings: (a) those involved in the control of body fluids; (b) those involved in the building of rigid structures to support the body; (c) those involved in chemical reaction in the body and as chemical constituents of the body.

Control of Body Fluids

The largest single component of the body is water, and this can be further divided into intracellular, intercellular, and intravessel water. The maintenance of a correct osmotic equilibrium and fluid pH in the body is essential for the correct transfer of nutrients and metabolites across cell membranes and around the body, and hence essential for life. The most important electrolytes that are present in body fluids and serve to maintain the correct osmotic equilibrium are sodium, potassium, chloride, bicarbonate, and phosphate ions.

The electrolyte composition of the intracellular fluid is quite different from the intercellular and intravessel (extracellular) fluid, the main difference being the relatively high amount of potassium and phosphate ions and the near absence of sodium and chloride ions in the former. However, the situation is not static, and whenever muscle or nerve is active there is a compensatory change in the electrolyte equilibrium at the cell membrane with sodium entering the cell and potassium leaving it. During recovery this process reverses itself.

The total amount of the main extracellular electrolytes sodium and chloride in an adult man is about 175 grams. These are readily lost from the body in two ways: in urine and in sweat. For example, in a temperate climate the daily amount of salt needed by an adult not doing active work is about 4 grams. The whole of this may, however, be lost in sweat during a few hours of exercise in warm conditions, and people usually have to eat between 5 and 20 grams of common salt per day with their food to maintain equilibrium. If the salt consumption is too low the amount of extracellular water is lowered in order to maintain osmotic equilibrium, and in extreme cases this can lead to serious disorders in the body.

About 98% of the main intracellular electrolyte potassium is present within cells, and about two-thirds of this amount is present in muscle cells. The capacity of the body for binding potassium is therefore related to the muscle mass of the body, and the body of a more muscular man tends to contain more potassium than that of a less muscular man. Potassium is also lost in urine but is not lost to an appreciable extent in sweat. The daily need for potassium is therefore low in relation to sodium at about 2–6 grams per day. This, coupled with the fact that most common foods contain appreciable amounts of potassium, means that there are seldom, if ever, cases of potassium deficiency or a need for the addition of potassium to a normal mixed diet.

B

TABLE 6. Dietary sources and biological function of minerals

Mineral	Recognised biological functions	Sources
MACRO MINERALS		
Calcium	Bones and teeth. Blood clotting. Nerve tissue regulation. Cardiac control. Regulation of muscular activity. Regulation of body fluids.	Milk and milk products. Fish and shellfish. Eggs. Fortified bread.
Chlorine	Associated with sodium and potassium in regulation of body fluids. Gastric secretion	Common salt and foods with added salt.
Magnesium	Component of soft tissue and bone. Correct Ca/Mg balance required for skeletal and nerve tissue function. Essential activator of phosphate transferring enzymes.	Widespread.
Phosphorus	Related to calcium in bones and teeth. General metabolism.	Widespread.
Potassium	Regulation of body fluids. Influences contractility of smooth skeletal and cardiac muscle. Affects excitability of nerve and muscle tissue. Intracellular metabolism.	Widespread.
Sodium	Regulation of body fluids.	Common salt and foods with added salt.
Sulphur	Component of essential nutrients, methionine, thiamin, and biotin. Sulphate also active in metabolism.	Proteins.
TRACE MINERALS*		
Chromium	Appears to be part of the glucose tolerance factor.	Widespread.
Cobalt	Component of vitamin B_{12}.	Liver. Meat.
Copper	Component of tyrosinase (polyphenol oxidase) and other enzymes. Inter-related with iron.	Widespread.

TABLE 6 (*continued*)

Fluorine	Prophylactic action on bones and teeth	Seafish. Water supplies. Tea.
Iodine	Formation of thyroid hormone which regulates metabolic rate.	Fish. Eggs. Cereals. Iodised salt.
Iron	Oxygen transport. Cellular respiration.	Liver. Meat. Shellfish. Nuts. Leafy vegetables. Fortified bread, and other cereal products.
Manganese	Activator of enzymes important to general metabolism, e.g. phosphatases, arginase.	Widespread. Normal diet contains approx. 4 mg/day.
Molybdenum	Component of two enzymes: xanthene oxidase and aldehyde oxidase. Possible catalytic role in fatty acid oxidation.	Widespread.
Nickel	Unconfirmed.	Widespread.
Selenium	Prophylactic action against liver damage and other disorders demonstrated in animals.	Widespread.
Silicon	Unconfirmed.	Widespread.
Tin	Unconfirmed.	Widespread.
Vanadium	Control of plasma and tissue cholesterol concentration. Phospholipid oxidation and autoacetyl CoA-deacylase activity.	Unknown.
Zinc	Component of metabolic enzymes, e.g. carbonic anhydrase, alkaline phosphatases, etc.	Widespread. Shellfish. Liver. Beans. Beef.

* Others suggested as possibly essential include arsenic, barium, bromine, cadmium, and strontium.

Building of Rigid Structures to Support the Body

The more important minerals in this group are calcium, phosphorus, and magnesium. These three elements in appropriate amounts are essential for the correct formation of bones and teeth. In the case of calcium, 99% of the total amount of calcium (about 1000–1200 g in an adult) occurs in bones and teeth. Although phosphorus is spread more widely in the body for general metabolism in the form of inorganic phosphates either free or coupled to organic molecules, about three-quarters of all phosphates (expressed as P) – about 600–700 g – is also present in bones and teeth. These two elements, together with much smaller amounts of magnesium (20–30 g), form a crystal lattice which is largely responsible for the rigidity and strength of bones and teeth. The weight ratio of calcium to phosphorus in bones is about 1.5:1, and this, coupled with the observation that excess calcium in diet diminishes the absorption of phosphates from food, suggests that 1.5:1 is a suitable ratio of intake of these two minerals from diet.

Of the rigid structure building elements, phosphorus and magnesium are present in nearly all foods in sufficient amounts for the chances of dietary deficiency to be remote. Calcium, on the other hand, is less widely distributed in foods in useful amounts, and dietary requirements depend more heavily on a narrow range of foods, particularly milk and milk products, fish when the bones are also eaten (e.g. sardines and canned salmon), and bread if it is fortified. In addition, the amount of calcium that the body will absorb from the diet depends on several factors. One of the most important is the intake of an appropriate amount of vitamin D. Without this vitamin the absorption of calcium from the food decreases, and in serious cases correct bone formation is no longer possible. The disease resulting from this is known as rickets. Less serious factors, certainly if one is on a well mixed diet, are components like phytic acid (present in bran and wheat), oxalic acid (present amongst others in spinach and rhubarb), and free fatty acids which form insoluble salts with calcium and interfere with absorption.

Thus, of all the rigid structure building elements, calcium is the one where deficiency is a possibility, and this is particularly true for children, who are growing and forming new bones and teeth, and expectant and lactating mothers who have to provide extra calcium for the development of the bones in the foetus, or provide milk for the newly born infant.

Factors Involved in Chemical Reactions in the Body and as Chemical Constituents of the Body

Aside from the maintenance of fluid balance and the building of rigid structures, many elements are required in diet to play important roles in the growth, maintenance, and repair of body cells of which muscle, blood, liver, kidney, etc. are composed, and in the release of energy during metabolism.

More classic examples in this category are as follows: the role of phosphorus in general metabolism; the role of sulphur-containing amino acids in the synthesis of body protein; the presence of certain elements in essential nutrients of the body, e.g. sulphur in the vitamins thiamin and biotin and cobalt in vitamin B_{12}; the role of iron in oxygen transport and cellular respiration; the role of iodine in the formation of thyroid hormones which regulate metabolic rate; the roles of a number of trace elements such as copper, zinc, magnesium, and molybdenum which are components of enzymes which play important catalytic roles in general metabolism.

Reference to review articles on these elements will show that they have been studied in much detail by nutritionists and biochemists. Such studies have already revealed much about their mechanisms of action, and the great importance of an adequacy of these elements has been clearly demonstrated in animal husbandry where supplementation of diet with minerals and vitamins is commonplace, and in laboratory experiments on animals with prepared diets. Although deficiencies have been observed in small and particular human population groups, such occurrences are relatively rare, probably because the foods in most human diets, unlike animal diets, have come from a range of plants and animals raised from a wide variety of soils. The exceptions to this are the elements iron and iodine, where intake (as with calcium) is more heavily dependent on a smaller range of foodstuffs; for example, for iron the main sources are meat (particularly liver), shell-fish, nuts, and leafy vegetables. For iodine, the main sources are sea- and shell-fish and vegetables grown on iodine-rich soils. For these two elements, clinical symptoms of deficiency are observed in humans, e.g. in anaemia for iron and goitre for iodine, in certain sections of the population.

Iodine deficiency is generally a more localised problem brought about by low levels of iodine in water and soil and hence in the native vegetables. In such conditions the use of iodised salt has proved to be an easy and effective preventive measure in the absence of other complicating factors such as the presence of high levels of goitrogenic agents in locally consumed vegetables of the brassica family.

The problem of iron deficiency is more complicated and has been the subject of continuing controversy for a number of reasons. For example, the availability of iron for intestinal absorption varies considerably from one food to another, depending upon the oxidation state and chemical form of the iron. Absorption is generally rather low, only 5–10% being absorbed from the average diet. The efficiency of absorption of iron from the diet varies from person to person and may be affected by their clinical condition; for example, increased absorption has been associated with conditions of starvation or pregnancy. On the other hand, decreased absorption has been associated with intestinal disorders. Finally, iron deficiency is in many cases caused by abnormal blood loss rather than by dietary deficiency. In order to minimise the chances of dietary deficiency of iron, a number of countries fortify suitable foods such as cereal flours with iron. The efficacy of such measures is, however, still a matter of some debate, and in most cases where anaemia occurs it is generally agreed that therapeutic administration of iron salts is a quicker and more efficient solution to the problem than dietary manipulation.

General Conclusions

The development of rapid and sensitive spectroscopic methods of assay and the application of radio-isotope tracer techniques have in recent years made possible an enormous expansion of knowledge on the role of minerals in the vital functions of the body. This will no doubt increase the number of minerals that are shown to be essential in the sense that they have recognised biological functions, but from the standpoint of practical nutrition it is important that we take into consideration a number of other factors before deciding where scientifically derived fact concerning essential nutrients is likely to be translated into human nutritional problems for the majority of healthy people on an adequate mixed diet. Such factors include the levels and distribution of a particular mineral in foods, body requirements, food intake patterns, absorption and retention by the body, and clinical evidence. On the basis of this type of evidence the minerals that are most likely to be associated with deficiency problems are iron and calcium, and it is because of this that many countries pay special attention to the development of a recommended daily allowance (RDA) for these two minerals. A summary of the more important minerals, their recognised biological functions, and their main dietary sources is given in Table 6 (pp. 24-25).

2.2 Vitamins

The vitamins are a group of organic compounds, differing greatly in chemical composition, which play essential catalytic roles in the normal metabolism of other nutrients. They cannot be synthesised by the body and must be obtained from diet. Because their role is primarily catalytic, in contrast to the gross nutrients like protein, carbohydrate, and fat, vitamins are required in relatively small quantities. They are found in varying quantities in a wide variety of foods, but no single food contains them all in sufficient quantities to satisfy human requirements under normal conditions of food intake. Though diseases now known to be due to vitamin deficiency in diet have been recognised for several hundred years and empirical cures applied, e.g. scurvy has been cured by eating fresh fruit or vegetables since the early 17th century, it is only during the past 50 years that the vitamins have been isolated in pure form, their chemistry and biological function further elucidated, and human requirements more accurately established.

Traditionally the vitamins have been divided into two groups on the basis of their solubility characteristics: fat-soluble vitamins and water-soluble vitamins. A certain usefulness is derived from such a grouping since it aligns the vitamins according to certain common physiologic characteristics and problems associated with analysis. Thus the fat-soluble vitamins A, D, and E are usually found in and extracted from food, in association with lipids from which they need to be separated prior to analysis. They are absorbed along with dietary fat. They are not normally excreted in the urine; they tend to be stored in the body, and man, given these reserves, is not absolutely dependent on their day-to-day supply in diet. In contrast, water-soluble vitamins are not normally stored in appreciable amounts in the body and any excess is excreted in small quantities in the urine. A more constant dietary supply of these vitamins is therefore desirable to avoid their depletion.

2.2.1 FAT-SOLUBLE VITAMINS

Vitamin A

The parent compound vitamin A, found only in fats of animal origin, is known as retinol and is an unsaturated monohydric alcohol with the structural formula shown in Fig. 6. A related compound, found in fish and having a vitamin A potency 40% that of retinol, is dehydroretinol, sometimes called vitamin A_2 (Fig. 6). Retinol is not found in plants but is present as precursors (or provitamins) in the form of carotenoids, which are converted into retinol *in vivo*. These carotenoids include various isomers of α-, β-, and γ-carotene and cryptoxanthin; the structures of the most biologically active isomers of these compounds are shown in Fig. 6 along with their approximate biopotencies relative to retinol. Of the carotenoids β-carotene is the most abundant in foods, and since its vitamin A potency is greater than that of other carotenoids it is regarded as by far the most important nutritionally. The vitamin A content of foods is therefore usually expressed as the sum of retinol and β-carotene expressed in retinol equivalents (i.e. 1 International Unit of vitamin A $\equiv$ 0.3 μg of retinol or 0.6μg of β-carotene). Vitamin A is essential for the integrity of epithelial tissues and normal growth of epithelial cells. It also plays important roles in the development of bones and teeth and normal vision.

Being fat-soluble it is absorbed in the intestine with dietary lipids and passes to the liver where it is stored. Normally the amount present in liver increases with advancing age and in healthy adults reserves are sufficient to meet the requirements of one or more years. Deficiency is therefore less likely in adults than in children where reserves are lower. Good dietary sources of vitamin A or its precursors are liver, eggs, butterfat, fortified margarine, certain fish liver

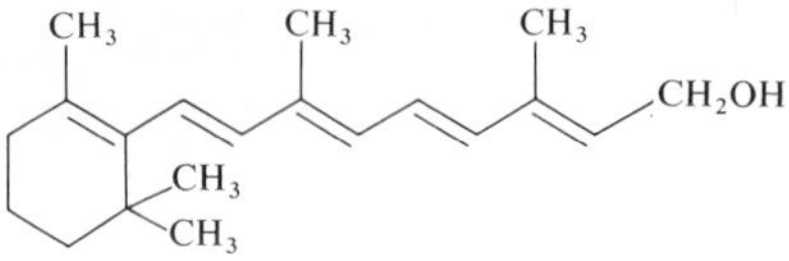

Vitamin A_1 (Retinol) 100

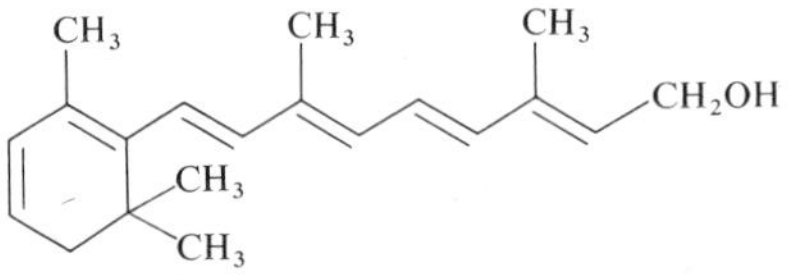

Vitamin A_2 (Dehydroretinol) 40

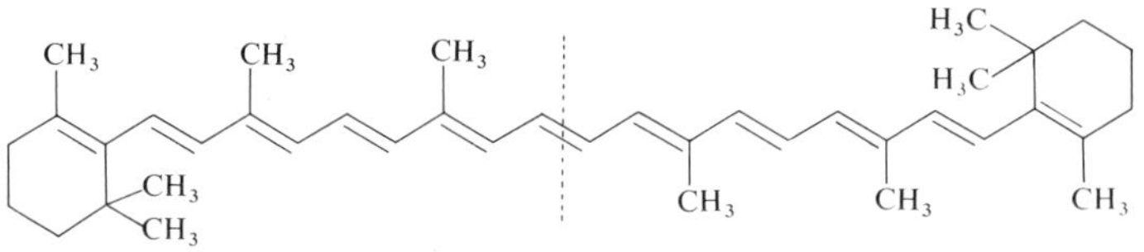

β-Carotene 50

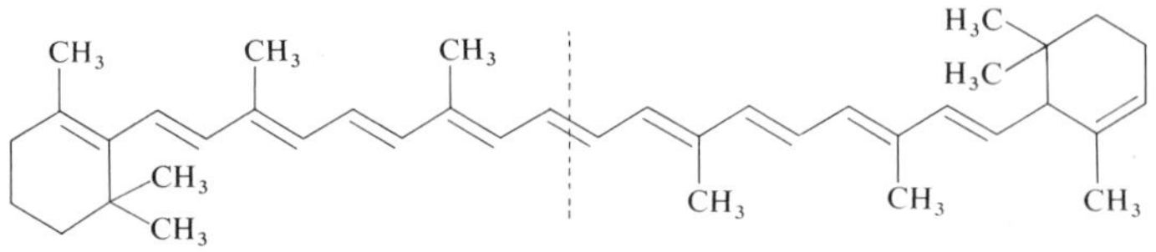

α-Carotene 27

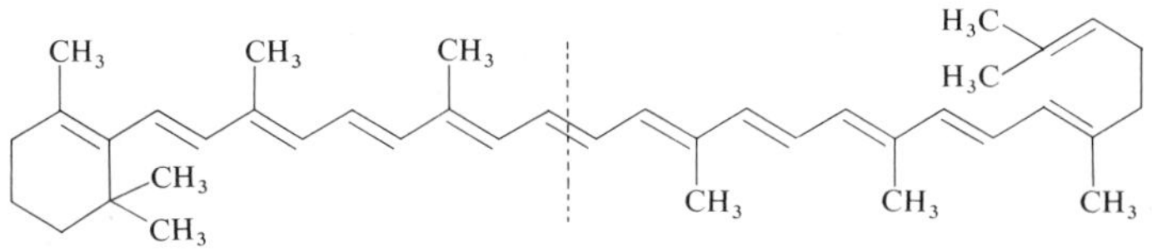

γ-Carotene 21

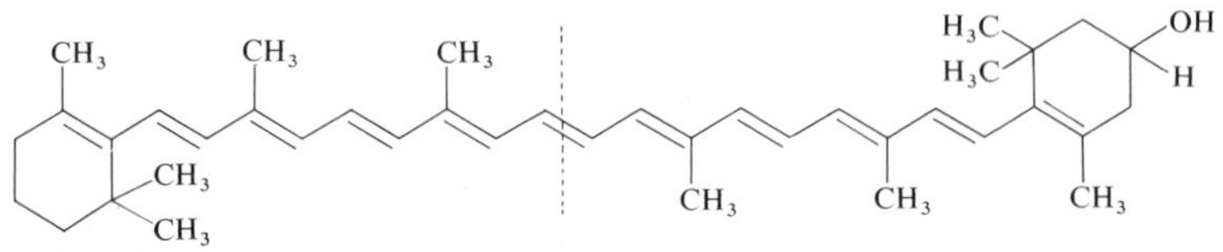

Cryptoxanthin 29

FIG. 6. Structure and biopotency of compounds with vitamin A activity. The values represent the approximate biopotency of the all-*trans* isomers by rat bioassay, relative to retinol (100).

oils, dark green or deep yellow leafy vegetables, and tomatoes. Clinical effects of vitamin A deficiency are usually seen only in people whose diet has been deficient in these foods over a long period of time or who are suffering from diseases that interfere with absorption of nutrients.

Since they are compounds containing conjugated double bonds, vitamin A and β-carotene are inherently unstable when pure, but in food they are stable to mild heat treatment during processing and cooking. At high temperatures or in sunlight and in the presence of oxygen, however, they are destroyed. They are oxidised by fat peroxides and are therefore destroyed under conditions that favour the oxidation of fats, including the presence of traces of copper and to a lesser extent iron. Thus, when synthetic vitamin A and β-carotene are added to food for enrichment, they require protection. This can be achieved by adding antioxidants. They are unstable to acid but can withstand boiling in strong alkali, as for example during the saponification used as a means of extraction before analysis. Losses (10–40%) of vitamin A and β-carotene can occur during processing and cooking particularly at high temperature, but being water-insoluble, losses by leaching action of cooking water are usually small. The food processor can use his skill to minimise such losses particularly by excluding oxygen.

Vitamin D Group

The D vitamins are substances related to sterols and are essential for the formation of normal bone. The most important compounds in this class are vitamin D_2 (ergocalciferol) and vitamin D_3 (cholecalciferol) and their precursors or provitamins ergosterol and 7-dehydrocholesterol from which they are derived through the action of ultraviolet light, as shown in Fig. 7.

Both forms of vitamin D have equal biological activity; the provitamins as such have none. Vitamin D_3 occurs as such in small amounts in some foods and is formed in the animal body by the action of ultraviolet light present in sunlight on the provitamin 7-dehydrocholesterol. In contrast vitamin D_2 does not occur naturally and is only formed in the body by ultraviolet irradiation of the provitamin ergosterol. Active vitamin D mixtures for fortifying food can be made by irradiation of suitable solutions of provitamins.

Vitamin D activates alkaline phosphatases from the kidneys, intestines, and bones which enhance net absorption of calcium and phosphorus essential to normal bone formation. Being fat-soluble it is absorbed in the intestine with dietary lipids. After absorption from the intestine or formation in the skin it is transported to the liver. In the liver the vitamins D_2 and D_3 are converted into the 25-OH derivatives after which in the kidneys the active 1,25-(OH)$_2$ derivatives are formed. These forms are involved in the formation of a calcium binding protein in the intestine which functions as a calcium carrier and hence an important regulator of calcium absorption from food.

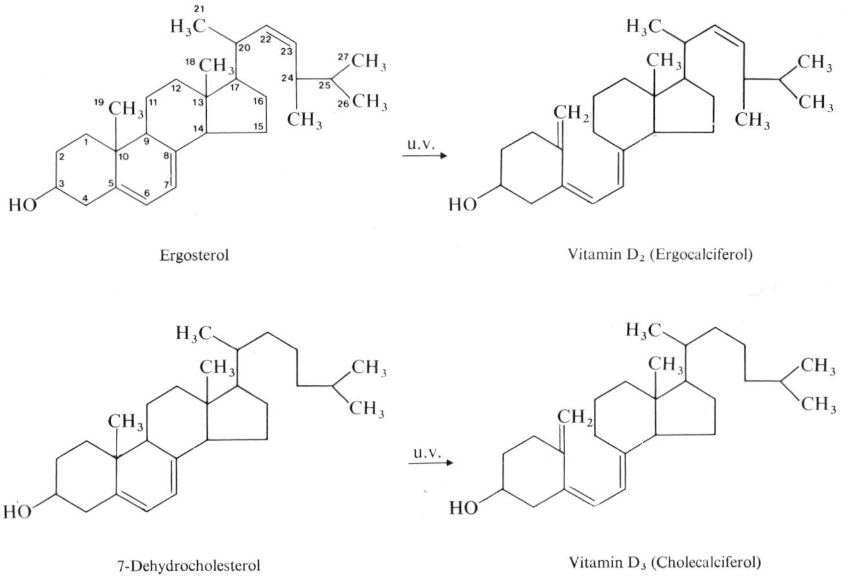

FIG. 7. Structures of vitamins D_2 and D_3 and their provitamins.

The body can store appreciable reserves of vitamin D, just as is the case with most other fat-soluble vitamins. These reserves probably account for the infrequency of vitamin D deficiency in adults; indeed, the healthy adult with adequate exposure to sunlight may have no dietary requirements for the vitamin. Dietary deficiencies can, however, occur in children whose reserves are smaller and in elderly housebound people exposed to little sunlight.

Most common food contains a small or negligible amount of vitamin D. The best natural sources are salt-water fish and their oils, liver, egg yolks, and, in summer, cow's milk. Because vitamin D is present in natural foods in small but variable amounts, some suitable foods are quite commonly fortified, e.g. fresh and particularly condensed milk which is widely used as a base for instant feeding formulations, some brands of butter, and most brands of margarine.

From very limited data available it appears that vitamin D in food is more stable than vitamin A.

Vitamin E

Vitamin E is a group name which includes a number of closely related biologically active compounds known chemically as tocopherols. This group is further subdivided into two general groups of compounds which have the trivial names tocols and tocotrienols. The tocotrienols differ from the tocols in that they possess three double bonds in their isoprenoid side-chain. The structures of tocol and tocotrienol are shown in Fig. 8.

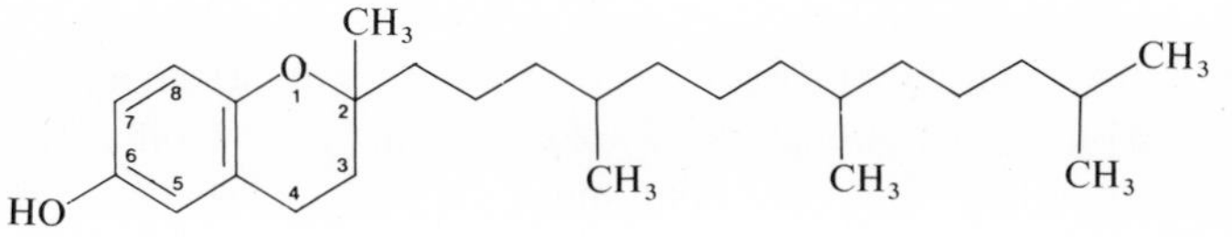

Tocol

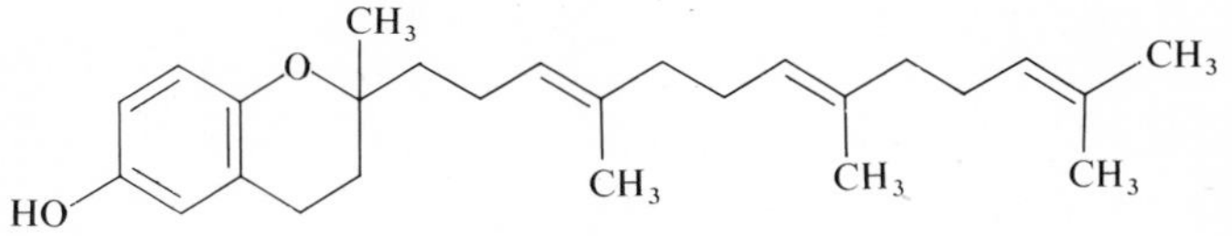

Tocotrienol

FIG. 8. Structures of compounds with vitamin E activity.

The individual tocopherols result from the substitution of one or more methyl groups on the tocol or tocotrienol ring system at the 5-, 7-, or 8-positions. Because of the presence of one or more asymmetric carbon atoms or double bonds a number of optical isomers and stereoisomers are possible. At least eight naturally occurring tocopherols have been isolated and others synthesised. Naturally occurring tocopherols are all in the optically active D-form. Synthetic compounds are of the D,L (racemic) form. Those natural tocopherols positively identified are the following:

Substitution	Tocopherol	Tocotrienol
8-Methyl	δ-Tocopherol (δT)	δ-Tocotrienol (δ₃T)
5,8-Dimethyl	β-Tocopherol (βT)	β-Tocotrienol (β₃T)
7,8-Dimethyl	γ-Tocopherol (γT)	γ-Tocotrienol (γ₃T)
5,7,8-Trimethyl	α-Tocopherol (αT)	α-Tocotrienol (α₃T)

The large number of tocopherols and their chemical similarity has resulted in great difficulties in the determination of the biopotencies of the various compounds. α-Tocopherol is the most active and widely distributed of the naturally occurring tocopherols and tocotrienols, followed by α-tocotrienol and β-tocopherol which have potencies approximately 30% that of α-tocopherol. Others, γ- and δ-tocopherol and the other tocotrienols, have only approximately 10% of the activity of α-tocopherol. Synthetic tocopherols show a similar order of biopotencies.

Vitamin E is necessary for normal reproduction in many animal species. It is also thought to be concerned in normal tissue respiration and to act as a regulator of phosphorylation reactions and metabolism of the cell nucleus. It is a powerful natural antioxidant, capable of being stored in food and body fat where it preserves easily oxidised materials such as unsaturated lipids and vitamin A. The richest dietary sources of vitamin E are the cereal seed oils (e.g. wheat germ oil). Other good sources are soyabean oil, corn oil, cottonseed oil, and margarine. Eggs, butter, and all meats (especially liver) contain some α-tocopherol.

Vitamin E is generally stable to heat, acids, and alkalis in the absence of oxygen or oxidising agents, but unstable to alkali in the presence of oxygen and oxidising agents. It is stable to visible light but destroyed by ultraviolet light.

Vitamin K

Numerous naphthoquinone compounds with vitamin K activity are known, the simplest and biologically most potent being synthetic vitamin K known as menadione or vitamin K_3. There are two major naturally occurring compounds related to menadione which have vitamin K activity. Their structures are shown in Fig. 9. Vitamin K is necessary for normal formation of prothrombin in the liver. Prothrombin is a glycoprotein present in blood plasma and necessary for normal clotting of blood.

Vitamin K is found in plant foods, particularly fresh dark green vegetables

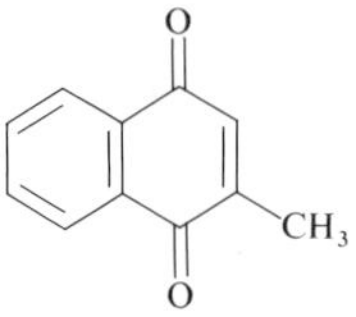

Vitamin K_3 (2-Methyl-1,4-naphthoquinone)

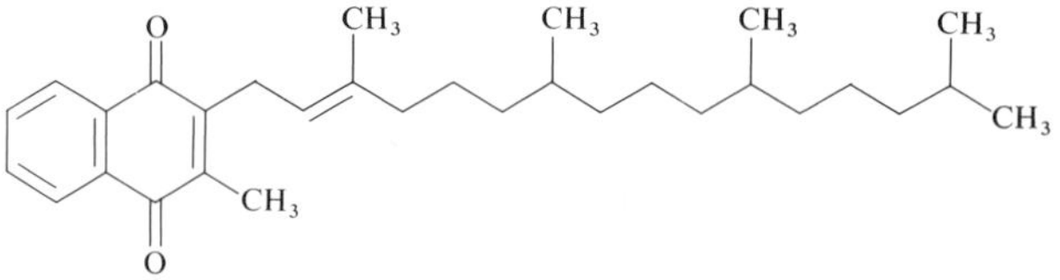

Vitamin K_1 (2-Methyl-3-phytyl-1,4-naphthoquinone)

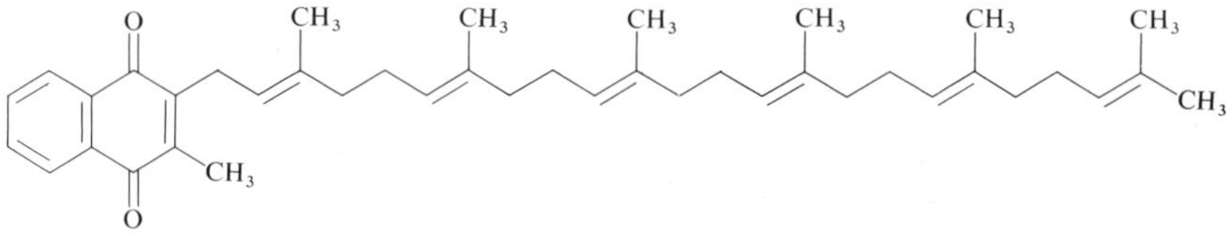

Vitamin K_2 (2-Methyl-3-difarnesyl-1,4-naphthoquinone)

FIG. 9. Structures of compounds with vitamin K activity.

such as kale or spinach. Foods of animal origin are generally poor sources unless they have undergone extensive bacterial putrefaction. Man's most important source of vitamin K appears to be his own intestinal bacterial flora, which provide him with adequate amounts under normal conditions of health.

The vitamin K group of compounds are generally stable to heat but are readily oxidised and labile to alkali, strong acids, and light.

2.2.2 WATER-SOLUBLE VITAMINS

Vitamin B_1 (Thiamin)
The structure of thiamin hydrochloride is shown in Fig. 10. Thiamin is essential for the maintenance of normal digestion and hence for good appetite and growth. The pyrophosphoric ester of thiamin is the coenzyme of carboxylase the enzyme concerned with the decarboxylation and oxidation of pyruvic acid which is of primary importance in the metabolism of carbohydrates. Thiamin requirements of the body are thus related to the level of carbohydrate in the diet. The amount of thiamin stored in the body is small, amounting in all to about 25 mg which allows for only a few weeks of normal functioning on body reserves. The body has no means of storing excess, so no benefit derives from taking large doses; the excess is lost in the urine. The food sources of thiamin are numerous, but comparatively few foods supply it in concentrated amounts. The richest sources are lean meats (particularly pork, and the offals, liver, and kidney), yeast, eggs, whole cereals, nuts, and rice. The milling of cereals and polishing of rice results in serious losses of thiamin. For this reason white flour is often fortified with thiamin.

Thiamin is comparatively stable to dry heat, much less stable to moist heat. Destruction under aqueous conditions is retarded in acid media (stable up to 120°C) and accelerated in alkaline media. Even the slight alkalinity of natural waters is sufficient to increase losses, and the use for example of alkaline baking powders can result in almost total loss of thiamin from flour. Thiamin is labile to oxidation and is destroyed by sulphites.

Since thiamin is water-soluble, large losses can occur from food by leaching, as is the case with all water-soluble vitamins, and losses are greatest with finely divided food immersed in large volumes of liquid. Thiamin is thus one of the

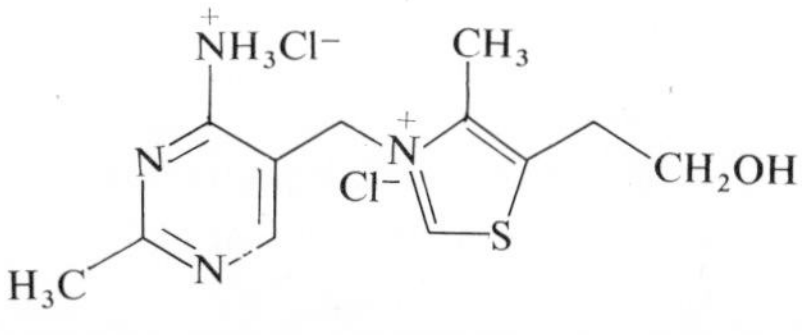

FIG. 10. Structure of thiamin hydrochloride.

most vulnerable vitamins, and considerable losses (up to 75%) can occur during processing or cooking if precautions are not taken.

Vitamin B_2 (Riboflavin)

The structure of riboflavin is shown in Fig. 11. In plant and animal tissues riboflavin is linked with phosphoric acid to give a mononucleotide. This is in turn combined with certain proteins to form catalytically active flavoproteins. These flavoprotein systems, in conjunction with the more complex flavin adenine dinucleotide (FAD), are capable of reversible oxidation and reduction and play a vital role in tissue oxidations essential to normal growth, such as those involved in the citric acid cycle. In view of the importance of riboflavin in cell respiration it remains a mystery as to how the clinical changes attributed to its deficiency in diet are minor and do not by themselves threaten life. One suggested explanation is that intestinal bacteria can synthesise the vitamin in small amounts, thus preventing an acute clinical deficiency picture from developing.

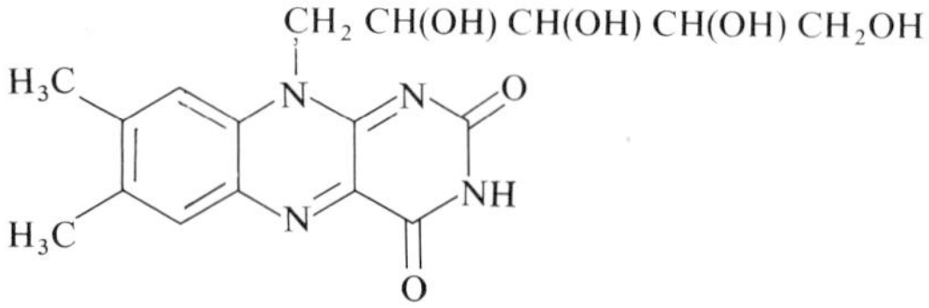

$$CH_2\ CH(OH)\ CH(OH)\ CH(OH)\ CH_2OH$$

FIG. 11. Structure of riboflavin.

The best sources of riboflavin in diet are liver, milk and dairy products, eggs, yeast extracts, and green leafy vegetables. It differs from other components of the B-group in that it occurs in higher amounts in dairy products but in relatively smaller amounts in cereal products.

Riboflavin is stable to oxygen and acid conditions but is readily decomposed by heat under alkaline conditions. It is also destroyed by ultraviolet light, and if foods, especially milk, are left exposed to sunshine large losses may occur. Heat alone is not harmful to the vitamin, and this coupled with its slight solubility in water serves to minimise losses under most cooking conditions although losses may be appreciable when water in which green vegetables have been boiled is discarded.

Niacin (Nicotinic acid, Nicotinamide)

In animal-derived foodstuffs almost all of the vitamin is found as nicotinamide either bound into nucleotides in living cells or as free nicotinamide in dead tissue and derived products such as carcase meat and milk. In plant-derived foodstuffs such as cereals and vegetables, variable amounts of both

Nicotinic acid   Nicotinamide

FIG. 12. Structures of nicotinic acid and nicotinamide.

nicotinamide and nicotinic acid are found depending upon the plant species and in free or bound form. The structure of free nicotinic acid and that of nicotinamide which is the physiologically active form are shown in Fig. 12. In humans the acid is readily converted into the amide.

The essential activity of nicotinamide is as a component of the nicotinamide adenine dinucleotide molecule (NAD). This molecule, like FAD, is capable of reversible oxidation and reduction and serves as an important catalyst in tissue oxidations.

The richest sources of niacin are yeasts, meats (especially organ meats such as liver and kidney), and cereal brans and germs into which the vitamin tends to be concentrated. It is also widely found in other foods but at relatively lower levels. The human body is not, however, entirely dependent on dietary sources of nicotinic acid to maintain its content of NAD, since tryptophan may serve as a precursor of NAD. (Approx. 60 mg of tryptophan $\equiv$ 1 mg of niacin.) The nicotinic acid equivalent (i.e. nicotinic acid + tryptophan content) of food is thus a better measure than nicotinic acid content alone, particularly for high-protein foods such as meat, eggs, milk, etc.

Nicotinic acid is one of the most stable of the water-soluble vitamins, being resistant to heat, light, oxidation, acids, and alkalis. Cooking causes little actual destruction of the vitamin, but appreciable amounts (e.g. 10–25%) may be lost in cooking liquors if these are discarded.

Vitamin B$_6$ (Pyridoxine and Related Compounds)
There exist three related compounds with potential vitamin B$_6$ activity: pyridoxine, pyridoxamine, and pyridoxal. Their structures are shown in Fig. 13. All three compounds occur in natural materials in the free form and in

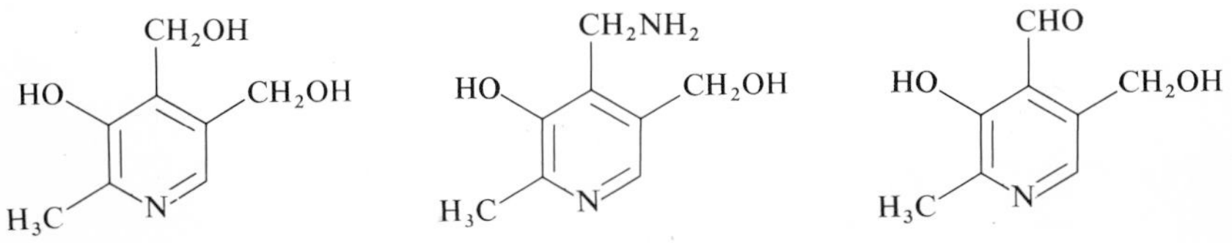

Pyridoxine   Pyridoxamine   Pyridoxal

FIG. 13. Structures of compounds with vitamin B$_6$ activity.

coenzyme form phosphorylated in the 5-position and/or bound to protein. In cereals, vegetables, and fruit, pyridoxine is the predominant form. In meats and milk, pyridoxamine and pyridoxal predominate. The vitamin as pyridoxal phosphate or pyridoxamine phosphate is involved in the synthesis and/or catabolism of all amino acids, e.g. in the transamination of the keto analogues of essential amino acids, the interconversion of glycine and serine, and the decarboxylation of many amino acids such as in the conversion of cysteine into taurine. Amongst a number of other important catalytic roles it is involved in carbohydrate metabolism and has been implicated in unsaturated fatty acid synthesis and the metabolism of nervous tissue.

Information on the vitamin B_6 content of foods is as yet incomplete but some of the recognised sources are brewer's yeast, liver, kidney, muscle meats, fish, whole grain cereals, rice, and potatoes. Compounds of the vitamin B_6 group, as with thiamin, are much less stable to moist heat than to dry heat. They are also sensitive to ultraviolet light and oxidation. Being water-soluble, B_6 group vitamins are lost in cooking liquors if these are discarded.

Vitamin B_{12}

The complex structure of vitamin B_{12} shown in Fig. 14 took many years to elucidate. Vitamin B_{12} is thought to participate in the transfer of single-carbon

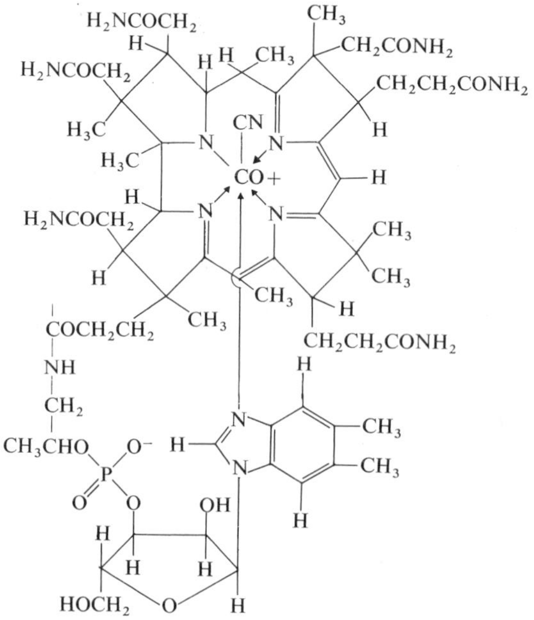

FIG. 14. Structure of vitamin B_{12}.

intermediates, especially methyl groups. In this role it has been linked to the function of folic acid and more particularly tetrahydrofolic acid (THF) which acts as carrier of one-carbon radicals that can readily be detached and incorporated into other molecules. Thus THF—CH_3 and homocysteine yield THF and the essential amino-acid methionine, and vitamin B_{12} is an essential catalyst in this reaction. Vitamin B_{12} also takes part in the formation of pyrimidine bases and in purine metabolism and is thereby involved in the synthesis of nucleic acids and nucleoproteins and in the formation of normal blood cells. Though the precise mechanisms of these processes are not fully understood, clearly the vitamin acts as a vital catalyst in some essential metabolic processes, since the amounts needed to transform a patient dying of anaemia into a healthy person are extremely small (i.e. 1 μg/day). Unlike most other water-soluble vitamins, vitamin B_{12} can apparently be stored in the body either in the blood conjugated with serum proteins or in the liver. Urinary excretion appears to be independent of dietary intake, and the amounts stored in the liver in the healthy well-fed adult are sufficient to meet the daily requirements for more than 1 year.

Vitamin B_{12} is found almost exclusively in foods of animal origin. Liver and kidney are excellent sources, muscle meat and fish supply it in moderate amounts, whole milk in lesser quantities, and most cereals are very poor sources. Vitamin B_{12}, being stable to heat at neutral pH but unstable in dilute acid or alkali, is one of the more labile water-soluble vitamins. It is also unstable to light and oxidising agents. Though cooking in itself may not destroy the vitamin, considerable losses can occur if cooking liquors are discarded.

Folic Acid

The parent compound, also known as pteroylglutamic acid, has the structure shown in Fig. 15. Folic acid itself is a synthetic product and not usually found in nature. In the natural forms usually found in food it is conjugated with up to six molecules of glutamic acid which are linked at the gamma carbon

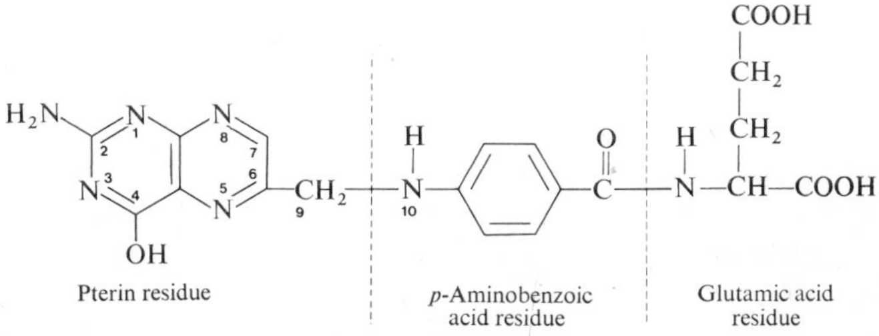

FIG. 15. Structure of folic acid.

positions. The physiologically active form of the vitamin is a reduction product 5,6,7,8-tetrahydrofolic acid (THF) in which formyl, hydroxymethyl, methyl, or formimino groups have been introduced on to the nitrogen at the 5- or 10-positions or linked across the 5- and 10-positions.

There are five known coenzyme forms of folic acid, collectively known as folate, and their major role is in the transfer of one-carbon units to appropriate metabolites in the synthesis of DNA, RNA, methionine, and serine. For example THF—CH_3 and homocysteine yield THF and methionine.

Despite the preponderance of knowledge about its essential metabolic role, there is a good deal of conflicting evidence about the dietary need for folic acid in the absence of antagonists, disease states, or deficiencies of other nutrients. Indeed it has been suggested by some authors that man, like the rat, may be able to derive enough folate from bacterial synthesis to meet the daily requirements.

Folic acid is widely distributed in food. Fresh green leafy vegetables, yeast, and liver are the best sources of the vitamin but useful amounts may also be found in some red meats, whole cereals, beans, nuts, certain fish and crustacea, and other vegetables. Milk and fruits are generally poorer sources of the vitamin.

Folic acid is sparingly soluble in water, and stable to heat in acid. When heated in neutral or alkaline media, however, it undergoes fairly rapid decomposition. Because of this, considerable losses can occur during some types of cooking.

Pantothenic Acid

The structure of pantothenic acid is shown in Fig. 16. Pantothenic acid is a constituent of coenzyme A and is present in all living matter. Its name is

$$HOCH_2-\overset{\overset{\displaystyle CH_3}{|}}{\underset{\underset{\displaystyle CH_3}{|}}{C}}-CH(OH)\ CO\ NH\ CH_2\ CH_2\ COOH$$

FIG. 16. Structure of pantothenic acid.

derived from the Greek word meaning 'from everywhere'. The distribution of the vitamin in food is so widespread (the richest sources being liver, kidney, yeast, egg yolk, and fresh vegetables) that deficiency is extremely unlikely to occur in man. Indeed, experimentalists in the field of pantothenic acid deficiency in humans have had to go to extraordinary lengths to produce diets that are free from the vitamin. Although stable in neutral solution the vitamin is readily decomposed by heat in the presence of acid or alkali to give pantoic acid and β-alanine.

Biotin

The structure of biotin is shown in Fig. 17. It forms part of enzyme systems, notably one that fixes CO_2 and incorporates it into the pathway of fatty acid synthesis.

Organ meats and yeast extracts are excellent sources of biotin. Pulses, nuts, mushrooms, chocolate, and some vegetables (e.g. cauliflower) are fair sources.

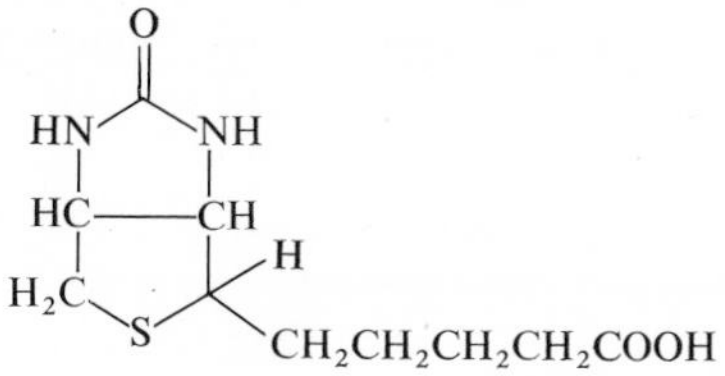

FIG. 17. Structure of biotin.

The human body probably utilises only a few micrograms of biotin daily, and it is very doubtful whether this needs to be included in diet. Yeasts and bacteria of many species either make or retain biotin, and it seems probable that man, even if he cannot make the vitamin in his own tissues, can obtain it from the numerous micro-organisms in his food and his intestines. Thus, provided that antinutrients are not persistently present in the diet, for example the protein avidin in raw egg white which combines with biotin rendering it inactive, man should have all the vitamin he needs. Biotin is sparingly soluble in cold water, stable to heat, and labile to oxidising agents and strong acid or alkali.

Other Vitamins of the B-Group

p-Aminobenzoic acid and inositol have been shown to be necessary dietary factors in certain animals. Choline is also sometimes classed with the vitamins of the B-group, being widely distributed in biological material either free or bound in the form of phospholipids. Their structures are shown in Fig. 18. They are undoubtedly physiologically active and important substances but there is no satisfactory evidence that human dietary deficiency ever occurs.

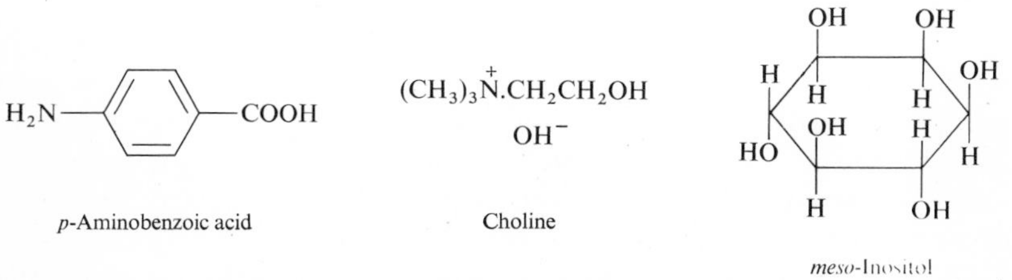

FIG. 18. Structures of other B-group vitamins.

Vitamin C

Ascorbic acid is the most active reducing agent known to occur naturally in living tissues and is easily and reversibly oxidised to dehydroascorbic acid (DHA) which is still physiologically active though less stable. Further conversion beyond the dehydroascorbic acid stage results in the irreversible formation of physiologically inactive diketogulonic acid. The structures of these compounds are shown in Fig. 19. Ascorbic acid is concerned with the integrity of connective tissue constituents, particularly collagen, and intercellular cement substance but its exact mode of action is not clear.

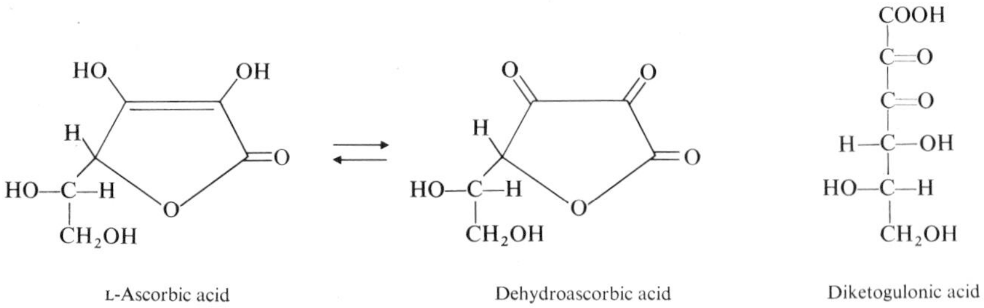

FIG. 19. Structures of compounds with vitamin C activity.

L-Ascorbic acid · Dehydroascorbic acid · Diketogulonic acid

The oxidised form of ascorbic acid (DHA) is thought to play an important role in the oxidation of the phenolic amino acids phenylalanine and tyrosine but, since only a small proportion of the vitamin is present as DHA in the body, this is probably not its chief activity. Although the precise influence of ascorbic acid on tissue metabolism is not fully understood in chemical terms, the effects of its deficiency on the structure of tissues have been documented for hundreds of years in cases of scurvy which is the classical manifestation of vitamin C deficiency.

Ascorbic acid has a more limited distribution than most other important water-soluble vitamins. Fruits and full-grown green vegetables are the richest sources. Root vegetables and potatoes contain smaller amounts, but the latter, although containing relatively lower amounts, is an important source, particularly in winter, because of its high daily consumption.

Though vitamin C is stable when dry, in aqueous solution it is highly labile and is easily oxidised especially in an alkaline medium and on exposure to heat, light, and traces of metals (e.g. copper and iron). It is more stable in acid solution particularly in the presence of metal chelating agents. Storage and processing can significantly reduce the amount of vitamin C contributed by food. Vitamin C is very sensitive to enzymic oxidation at room temperature, which can be induced by cutting or bruising, as well as chemical oxidation on heating. The vitamin is also very soluble in water, and processes such as washing, blanching, or cooking may lead to serious losses through the leaching action of excess water.

CHAPTER 3

Analysis of Nutrients in Food

3.1 Introduction

Although our understanding of the chemistry and biochemistry of food systems and the sophistication of our food technology has been developing rapidly over the past 30 years, the industrial food analyst's role has tended to be limited, until much more recently, to the use of large numbers of manual, proximate analyses of the main food components such as moisture, ash, fat, protein, and carbohydrate (although the latter was often in the past calculated by difference) and small numbers of specialised analyses based upon classical wet chemical procedures. The attraction of these methods lay in the fact that they were capable of being performed in a routine laboratory with fairly simple apparatus and relatively unskilled staff. Furthermore, in the age of more conventional food materials and simple food technology, such proximate and wet chemical analyses were probably sufficient for most quality control purposes.

Although these manual proximate and wet chemical analytical procedures still have considerable value in terms of giving general information about the gross make-up of a food product, it is important to emphasise that they have a number of shortcomings. For example, they are incapable of providing sufficient detail to satisfy the increasing interests of governments and consumers concerning the nutritional and safety aspects of food. They tend to be precise, not accurate, and in many cases lack the specificity to cope with the complexity of many modern food products. They are often, especially for the less routine analyses, time and labour consuming and increasingly unable to cope with the enormous diversity of products produced by modern food technology.

Fortunately, during the last 10–20 years a range of instrumental methods has been developed and perfected to overcome these shortcomings. For example, the techniques of gas–liquid, ion-exchange, and high performance liquid chromatography have increased enormously the analyst's ability to separate complex food mixtures. Spectroscopic and other modern detection systems have similarly increased both the sensitivity and specificity of methods, and mechanisation/automation has provided a means for dealing, at least in part, with the volume of data now required. This book is intended to reflect these modern developments but at the same time it recognises the widely differing operational conditions of individual analysts and hence the continuing need for and values of less sophisticated methods of analysis.

3.2 Sampling – Errors and Procedures

Food analysis is always subject to a number of errors which affect both its accuracy (its ability to give a 'true' answer) and its precision (its ability to reproduce an answer). These errors are of two main types as follows.

Indeterminate errors. The magnitude of these errors cannot usually be predicted, and it is essential that they are traced, explained, and eliminated. They are due to serious experimental mistakes or fluctuations in the environment. Typical examples would be mathematical errors in calculations or those caused by vibrations when using a very sensitive balance.

Determinate errors. These are inherent in the total analytical procedure used and may be operational (technique or reagents) or instrumental in nature. They may be 'random' in the sense that they follow the laws of chance and their frequency is placed about the mean result in a large number of experiments (i.e. they relate to the precision of the method), or 'systematic' which means that they are constant and in the same direction, e.g. those caused by faulty calibration, personal prejudice, and chemical interference (i.e. they relate to the accuracy of the method).

Three main steps are involved in the analytical determination of a food component, each of which may be subject to error: (1) the taking and preparation of a sample; (2) the working-up of a sample; and (3) the measurement of a feature of the worked-up sample. Clearly if the first step is seriously in error any subsequent steps may prove to be useless and sampling should therefore be looked upon as an important operation with its own set of problems and techniques for overcoming them.

The objective of sampling from a nutritional standpoint is to ensure that the sample taken for analysis is representative of a defined whole and also representative of what is to be eaten. There is no simple solution to this since the complexity of the problem is dependent upon the size and the nature of the defined whole, both of which are infinitely variable. For example, size may range from a ton of apples hanging in an orchard to a single apple on a plate, or nature may vary from heterogeneous particulate matter to a completely homogeneous liquid. As a consequence the complexity of the sampling problem may range from one that requires the application of statistical analysis techniques to one that can be resolved through common sense and experience. It is not the intention of the authors to provide any further detail on statistical analysis techniques in this text but merely to highlight the main problems associated with the taking and preparing of samples.

Outside operational and instrumental error, there are two main sources of error in the sampling area as follows.

The failure of the sample to represent the whole owing to inhomogeneity. This may occur at the primary sampling stage both from a difference in composition between individual food items of the same variety and maturity, and from the difference in composition between various parts of the same food item, for example, the variation in the vitamin content of a number of individually grown cabbages of the same variety and maturity and the variation in the vitamin content between the outer leaves and the heart of a single cabbage. It is essential that sufficient material is taken to compensate for these variations. It may also occur in the subsequent sub-sampling stages to obtain a quantity suitable for analysis owing, for example, to inadequate homogenisation or phase separation on standing.

In order to highlight sampling problems of this type the analyst will usually take more than one sample for analysis and observe the precision of his results. If the precision of the results is worse than he would have expected from the method using reference samples, then sampling problems are indicated and the sampling procedures should be re-examined.

Changes in composition of the material during sample preparation. Examples of this are the loss or absorption of moisture, the loss of volatile constituents, and the chemical or enzymic decomposition of sensitive components such as vitamins. These are related to the effects of handling procedures such as reduction in particle size by grinding, milling, chopping, or pulverisation, the degree of mixing by shaking or stirring, the need for drying to stabilise or make the sample easier to handle, and the effects of the length and conditions of storage prior to analysis.

Many of these procedures have been shown to cause undesirable changes in the composition of particular nutrients in particular foods. It is not the intention of the authors to catalogue these but to increase the reader's awareness of the general type of problem that may arise by giving a few examples, e.g. the contamination of the sample with minerals by mechanical erosion from grinders, the oxidation of sensitive components caused by aeration during blending, the loss of moisture or decomposition of components owing to the generation of excessive heat on mixing, the rapid enzyme changes that can occur during the crushing or drying of plant and animal tissues, and the deterioration of samples owing to microbial spoilage on storing.

The authors, in providing only one general procedure for sample preparation which they believe has fairly wide application and minimises the chances of any problems arising, recognise that in practice there will be a number of ways in which a particular food sample may be (or has to be) handled prior to analysis without problems arising, which will be governed by such factors as the availability of apparatus and the experience of the analyst with a particular food.

3.3 Analysis for Macronutrients

3.3.1 PROTEINS

The accepted standard method for the determination of protein in foods involves the complete digestion of the sample in hot concentrated acid and in the presence of an appropriate metal ion catalyst such as copper or mercury, to convert all the nitrogen in the nitrogenous materials in the sample into ammonium ion. Upon the addition of alkali to the digest ammonia is released which may then either be distilled out of the sample and determined by simple acid–base titration as in the classic Kjeldahl procedure or alternatively the ammonia can be reacted with an appropriate reagent such as phenol and sodium hypochlorite, to give a coloured derivative which can be measured spectrophotometrically.

Whilst these methods have great versatility, a proven record of reproducibility, and are invaluable to the analyst for routine purposes, they are not without their limitations. Firstly, it should be remembered that these methods do not measure protein directly but use an empirical factor to convert nitrogen into protein. This empirical factor varies from protein to protein (e.g. cereal protein $N \times 5.70$, milk protein $N \times 6.38$, general factor $N \times 6.25$) which may be a source of inaccuracy. Secondly, unless additional steps are taken to remove non-protein nitrogen from some samples before analysis, the methods may overestimate the amount of 'true protein' in the sample. Thirdly, and most important from a nutritional point of view, although these methods may give an accurate estimate of the quantity of protein in the sample, they can say nothing about its quality.

The area of protein quality assessment is a complex one in which biological tests are still the ultimate reference point. In such tests the nutritional value of proteins may be assessed by bioassay using the young rat as an assay organism.

The most widely used tests are Net Protein Utilisation (NPU) which measures the percentage of intake nitrogen retained for growth and maintenance, Protein Efficiency Ratio (PER) which measures the weight gain per gram of protein eaten, and Biological Value (BV) which measures the amount of absorbed nitrogen that is utilised for protein synthesis. Such tests enable food proteins to be ranked in order of nutritive value or compared with a reference protein such as casein.

Although biological methods are to be preferred, the analyst can gain some estimate of the quality of the protein via chemical tests which determine its amino acid composition. This is achieved by acid or, in some cases, alkaline hydrolysis of the protein in the sample followed by a separation and quantitation of the liberated amino acids using ion-exchange chromatography.

Having obtained the amino acid composition one can then compare the quantities of the essential amino acids in the protein with the quantities in a suitable reference protein and arrive at a measure of the maximum potential nutritional value of the protein under test which is commonly referred to as the 'chemical score'.

Ideally there should be a direct correlation between the chemical score and values determined by bioassay techniques. In practice this is seldom absolutely the case because other factors interfere. Nevertheless, amino acid analyses are a useful semi-quantitative guide to the nutritional value of proteins.

3.3.2 CARBOHYDRATES

The term carbohydrates embraces a broad spectrum of compounds ranging from simple mono- and di-saccharides to complex polysaccharides. The most common approach to the determination of the carbohydrate content of foods is by 'difference', that is by deducting the sum of the measured moisture, ash, protein, and fat from the total weight. There are a number of objections to this approach. Firstly, it is vulnerable to the inaccuracies associated with the determination of the other constituents; secondly, it does not take into account other minor components such as lignin which may be present. Thirdly, and most important from a nutritional point of view, it does not differentiate between carbohydrates that are available to man and those that are not.

The second and third of these objections can be overcome to a useful degree by applying two further methods. The first of these, the Clegg anthrone procedure, involves digesting the food sample with perchloric acid and determining solubilised starch and simple sugars colorimetrically, to give a rapid guide to the total available (or digestible) carbohydrate content. The second of these, the crude fibre method, involves the estimation of the insoluble material remaining from a food sample after boiling both in acid and then alkali or alternatively an acid detergent solution, which gives a similar guide to the non-available (or non-digestible) carbohydrate and lignin content. This information, when coupled with the data from approximate analysis, may then prove sufficient.

If more accurate information is still needed, this requires the measurement of the individual components (or groups) which make up the available carbohydrate fraction. Free sugars (including glucose, sucrose, lactose, maltose) can be collectively determined in an aqueous ethanolic extract of the food sample after hydrolysis by copper reduction methods. Starch can similarly be determined on the residue from the alcoholic extraction again after suitable hydrolysis (either chemical using hot mineral acid, or by an enzyme such as amyloglucosidase).

More detailed characterisation of individual free sugars present as such or in the hydrolysates from polysaccharide fractions may sometimes be useful, and this can be obtained by using specific enzyme tests or by using high performance liquid chromatography on ion-exchange supports.

3.3.3 FATS

The method used for the determination of fat in food depends on the type of lipid present, the other food components with which it is mixed, and the degree of accuracy and detail required from the analysis. Thus the analyst needing to establish the total fat content of, for example, a dry cereal-based ingredient in which the lipid fraction is composed mainly of triglyceride, will probably find that continuous extraction in a Soxhlet apparatus with ether or petroleum ether is quite satisfactory. On the other hand, when examining a wet milk-based dessert product containing in addition to triglyceride significant quantities of mono- and di-glycerides present as emulsifiers, as well as heat- and acid-sensitive sugars, he will probably require a quite different technique, e.g. the Roese–Gottlieb method in which pre-treatment of the sample with alcoholic ammonia is used to break up protein–fat interactions without destruction of carbohydrates prior to the extraction with solvent. Yet another method, which will probably be the most suitable for wet proteinaceous products not containing carbohydrates (which would decompose and interfere under the conditions of the test), is the Weibul technique in which the sample is digested in concentrated acid prior to solvent extraction.

There are thus many methods for fat determination which have been developed to suit a particular application (see notes on application in Methods) and few have general application. An exception to this is the method based on extraction with chloroform and methanol which is of proven use for extracting all the main classes of lipid from many foods. Though tedious for routine work, it has general application and is the method of choice when further information is required on the fat fraction. For example, the residue from a chloroform–methanol extract is used as the starting material for obtaining a complete fatty acid profile from the lipids by gas chromatography, which is useful both for identifying the type of fat and for determining essential/unsaturated fatty acid. It is also the starting material for determining individual classes of lipid, e.g. mono- and di-glycerides, cholesterol, phospholipids, by other chromatographic procedures, e.g. thin-layer chromatography. These are, however, seldom required in routine nutritional analysis.

Calorific Value
The calorific value of food is usually calculated from the amounts of protein, fat, and available carbohydrate that it contains, these amounts being derived by appropriate methods as described above. The amounts of protein, fat, and carbohydrate are then multiplied by factors representing the number of kilocalories produced by 1 gram of the materials in the body. The sum of these gives the calorific value of the food (see p. 18, and typical calculation on p. 239).

3.3.4 MOISTURE

The determination of moisture content is one of the most important and widely used measurements in the processing and testing of food products. Although simple in concept, in practice the accurate determination of moisture content is complicated by a number of factors which vary considerably from one food to another, e.g. the level of moisture in the sample, the ease with which that moisture can be removed (which depends upon the relative amounts of 'free' and 'bound' water), the rheology of the sample, the extent to which there might be a loss in weight from the sample through thermal decomposition or a gain in weight due to oxidation, and the extent to which there may be a loss of volatiles other than water from the sample.

As a result, a wide variety of basic methods and variations within each method (e.g. changes in temperatures, times, extraction solvents, sample preparation, etc.) have been developed and used for the determination of moisture content in foods. These may be classified under a number of headings as follows.

Methods that are based upon the removal of water from the food and its measurement by loss of weight or the amount of water separated. Air or vacuum oven drying at 70–140°C are considered to be direct and reliable methods provided that there is no thermal decomposition of the sample and water is the only volatile constituent removed. Distillation techniques, in which the volume of water released from the sample by continuous distillation with an immiscible solvent of appropriate boiling point (toluene 110°C to xylenes

$\sim 140^{\circ}$C) is measured in a specially calibrated receiver, are particularly useful for avoiding the difficulties associated with materials that contain appreciable amounts of volatiles other than water, but care still needs to be taken to avoid problems associated with thermal decomposition of sample components.

Provided that there are no serious interferences through decomposition and the presence of other volatiles, the simplicity and directness of these methods will tend to make them the methods of choice for routine work, particularly on fresh foods of higher moisture content (e.g. 60–95%), in a laboratory which has to deal with a wide range of samples.

Methods that depend upon the measurement of some physical property of the product which changes regularly with water content. These include electrical methods based upon conductivity, resistance, capacitance, and dielectric constant, techniques based upon nuclear magnetic resonance and infrared spectroscopy, and other physical techniques involving density, vapour pressure, and refractive index. The success of these physical procedures is dependent upon the physical property being much more sensitive to changes in water content than changes in other components over a wide enough range. These methods are not direct, and it is necessary to calibrate the process for each type of material against appropriate reference standards; also, few of them are precise over the full range of moisture levels found in fresh and processed foods (i.e. 5–95% of moisture).

Such methods are generally of most value in laboratories where there is a large and constant supply of the same materials for analysis which have intermediate moisture contents (e.g. 5–20%). For example, instruments that exploit electrical properties of the sample are commercially available for moisture determination in dehydrated fruits and vegetables and cereals.

Methods that depend upon the chemical reactivity of water. These include the generation of acetylene from calcium carbide or hydrogen from calcium hydride, the generation of heat on mixing with concentrated acid, the change in colour of cobalt chloride, and the conversion of iodine into iodide in Karl Fischer reagent.

Most of the methods have been devised particularly for the determination of small quantities of water present in organic fluids rather than in foods, the exception to this being the Karl Fischer method. In this method a food sample is first extracted with methanol to remove all the water. Karl Fischer reagent, which is prepared from iodine, sulphur dioxide, and pyridine in methanol solution, is then titrated into the sample extract and the end-point can be assessed either visually by the brown colour produced due to excess iodine or electrometrically. The reaction, which is near stoichiometric in the absence of interferences from certain aldehydes and ketones, may be represented as follows:

$$SO_2 + I_2 + 3C_5H_5N + H_2O \rightarrow 2C_5H_5N.HI + C_5H_5N \underset{O}{\overset{SO_2}{\diagdown | }}$$

$$\Bigg| CH_3OH$$

$$\downarrow \quad SO_4CH_3$$

$$C_5H_5N \underset{H}{\diagup }$$

Despite the rather exacting procedures that are required to calibrate the process and protect it from the ingress of moisture which limits the throughput of samples, many analysts would support the Karl Fischer method as a reference procedure, particularly for the determination of low levels of moisture in foods (i.e. $<10\%$) to which it is particularly well suited.

Faced with a widely differing range of foodstuffs, the analyst will often need to decide whether it is more important for him to have available an equally wide range of methods, to achieve the most accurate result, or whether it is adequate to apply a more limited number of methods, which in some cases will limit his accuracy but give good replication so that results are comparable and at the same time save him labour, time, and cost.

For this reason the authors, whilst recognising that there are a multitude of methods for moisture determination that could usefully be applied in particular circumstances, have chosen a limited number of general methods, e.g. air and vacuum oven drying, distillation, and the Karl Fischer method, which together should provide a sufficiently wide range of cover for most situations.

3.4 Analysis for Micronutrients

3.4.1 MINERALS

When determining the levels of inorganic nutrients in food, the first objective is to obtain the minerals in concentrated form separated from as many sources of interference as possible. This is invariably achieved by destruction of the organic food matrix by wet oxidation with, for example, concentrated nitric and sulphuric acids or dry ashing, the conditions being chosen so as to minimise losses and obtain the elements that require measuring in the most readily handled form for subsequent determination.

The second objective is the determination of individual elements. This may be achieved by electrochemical (polarographic, coulometric, conductometric, potentiometric, amperometric analysis), atomic spectroscopy (emission, absorption, fluorescence), colorimetric/spectrophotometric, chromatographic, complexometric, volumetric, and gravimetric techniques.

There is thus a wide choice of procedures for determining each element, and readers wishing to know more about mineral determination in general should consult the introductory references at the end of this manual as well as the current literature. For the purpose of this manual, methods have been chosen that have a history of simple and reliable application to food analysis. Total inorganic matter is determined by dry ashing of the food sample. The residual ash is then digested in hydrochloric acid, and the metal chlorides are taken up into an appropriate volume of dilute acid and determined using atomic absoprtion spectrophotometry. Alternatively, the sample may be wet digested prior to determination by atomic absorption spectrophotometry. Elements thus determined are calcium, magnesium, sodium, potassium, copper, zinc, iron, and manganese. Although atomic absorption is the method of choice for most metals, suitable flame photometric methods are also available for sodium and potassium. Total phosphorus is also conveniently determined by colorimetry on an appropriate extract of the ash or aliquot of wet digest. Chloride is determined coulometrically or titrimetrically on a direct extract of the foodstuff without recourse to ashing or digestion.

The remaining elements, molybdenum, iodine, fluorine, chromium, selenium, sulphur, and cobalt, are not normally required in routine nutritional survey work, and procedures are not therefore included in this book. Separate methods are, however, included for important compounds containing some of the elements. For example, sulphur is determined as cysteine, methionine, and thiamin, and cobalt as vitamin B_{12}, since these are the active principles rather than the elements themselves.

3.4.2 VITAMINS

Many of the vitamins are labile components (see section on vitamins), and the amounts being measured in the final analytical procedure often amount to only a few micrograms. Additional precautions must therefore be taken during the handling and subsequent analysis of samples, to ensure that factors such as pH, heat, light, air, or oxygen do not destroy vitamins and that recoveries of vitamins from food are good.

Fat-Soluble Vitamins
The fat-soluble vitamins discussed are vitamins A, D, and E. Other fat-soluble vitamins, e.g. vitamin K, are not discussed in this section of the book since neither data on dietary requirements nor suitable analytical procedures are available.

There are two major classes of method available to the analyst for the determination of fat-soluble vitamins: biological and chemical. Of the two, the biological methods have the advantage of supplying an accurate estimate of biopotency without elaborate separation of components. On the other hand, they suffer the disadvantages of long analysis time (e.g. weeks), and the need for animal facilities. It is therefore not surprising that analysts have searched for, and now largely rely on, alternative chemical methods. Chemical methods all require preliminary separation of the vitamins from fat. This is achieved by saponification and extraction of the unsaponifiable function, which contains the vitamins. Many variations in procedure have been suggested for determination of vitamin A and provitamin A but it is now generally recognised that chromatography of the unsaponifiable fraction from a food on columns of alumina and magnesia is a reliable and widely applicable means of separating vitamin A and carotene. The separated components are then measured spectrophotometrically or fluorimetrically.

Two versions of the recommended procedure are given in this book. The principle in each is the same, but one uses classical adsorption column chromatography techniques and the other the more recently developed high performance liquid chromatographic techniques.

Some progress has recently been made with the development of chemical methods of analysis for vitamin D. The levels of the vitamin in both natural and fortified foodstuffs are, however, extremely low, ranging from 2 $\mu g/g$ in the richest sources, such as cod liver oil, to 0.001 $\mu g/g$ in cow's milk. It is therefore not surprising that until quite recently biological assay procedures involving rats or chickens were the only methods capable of estimating these low levels.

The main problems associated with chemical methods of analysis for vitamin D in foodstuffs are the removal of interfering components such as cholesterol, vitamin A, and vitamin E, which are invariably present in gross excess, by appropriate clean-up procedures (see Table 7) and the sensitivity and specificity of the determinative step.

A number of determinative techniques have been suggested in the past, for example colorimetry and ultraviolet and infrared spectroscopy, but none is satisfactory for foodstuffs. More recently gas–liquid chromatography of electron-capture sensitive derivatives of vitamin D has been studied and appears to offer a method capable of reliably determining the vitamin at the 0.1 $\mu g/g$ level in certain foods. Methods are currently being developed for determining the vitamin at even lower levels for more general application using

TABLE 7. Ratios of relevant components in some foodstuffs.

Foodstuff	Ratio relative to vitamin D on a weight basis			
	D	A	E	Cholesterol
Whole milk	1	1500	5000	600 000
Ox liver	1	6000	1400	300 000
Whole egg	1	75	500	80 000
Cod liver oil	1	120	60	10 000

improved clean-up procedures and more specific detection systems, for example gas chromatography of derivatives coupled with mass spectrometry, but such methods are not included in this book because they require very specialised equipment that is available only in a limited number of laboratories.

There are a large number of chemical methods available for the determination of vitamin E, but few have general application to food analysis. Separation of the tocopherols from the unsaponifable residue may be accomplished by a number of techniques, e.g. column chromatography, one- and two-dimensional paper or thin-layer chromatography, and gas–liquid chromatography. With the exception of gas–liquid chromatography the separated materials are commonly measured spectrophotometrically. Currently the preferred methods for vitamin E in food are based on one- or two-dimensional thin-layer chromatography to separate the tocopherols which are then determined by spectrophotometry.

Water-soluble vitamins

The water-soluble vitamins for which methods of analysis are given in this manual are those of the B-group, B_1, B_2, niacin, B_6, B_{12}, and vitamin C. Other water-soluble vitamins, e.g. folic acid, pantothenic acid, biotin, etc., are not discussed in any detail in this book either because suitable analytical procedures have not yet been fully established (folic acid) or dietary requirements and deficiencies have not been clearly defined (pantothenic acid, biotin).

B-Group Vitamins

There are again two major classes of method available to the analyst: microbiological and chemical. The microbiological methods have the advantage of giving an accurate measure of biopotency, the growth response of the particular micro-organisms being specific to a given vitamin. Satisfactory microbiological methods are described for the vitamins riboflavin, niacin, B_6, and B_{12}.

Microbiological assay procedures, although less demanding than animal assay procedures, still require special equipment and training for staff and are rather tedious and time-consuming. Analytical chemists have again therefore searched in this area for alternative and more rapid chemical procedures. Many chemical methods are now available for the analysis of B-group vitamins but few are suitable for the determination of all of the vitamins in a wide enough range of food products. There are, however, satisfactory chemical methods for the vitamins thiamin, riboflavin, and niacin.

Chemical methods require a sample pre-treatment procedure to liberate the B-group vitamins which are coupled with proteins, phosphate groupings, etc.

For the vitamins thiamin, riboflavin, and niacin this is accomplished by chemical hydrolysis, followed by enzymic digestion of the sample with suitable proteases and phosphatases to release the vitamins.

Clean-up of the food extract is then achieved using either classical or high performance liquid chromatography on silica. The preferred determinative technique for both vitamins is fluorescence spectroscopy, thiamin as thiochrome and riboflavin directly. Further research is continuing towards the developments of chemical procedures for other B-group vitamins (B_6, B_{12}) but as such methods cannot as yet be shown to compare favourably with existing microbiological procedures they are not included in this book.

Vitamin C

The best-known procedure for the determination of vitamin C is a redox titration technique with 2,6-dichlorophenol-indophenol. Certain compounds such as thiols and reductones may interfere with the method if adequate precautions are not taken. An additional and very specific method for vitamin C is based on oxidation of ascorbic acid to dehydroascorbic acid and subsequent reaction with *o*-phenylenediamine to form a fluorescent quinoxaline derivative. This method allows the analyst to measure both ascorbic acid and dehydroascorbic acid.

C

CHAPTER 4

Recommended Intake of Nutrients and Interpretation of Nutritional Data

4.1 Recommended Intake of Nutrients

Tables of recommended daily intakes (RDI) have been prepared by several official national and international bodies. Such tables serve a number of useful purposes. Firstly, they provide a standard against which the diets of different sections of the community or a particular socioeconomic class can be measured. Secondly, they provide a guide for caterers and dieticians planning large-scale feeding, for example in hospitals, schools, and canteens. Thirdly, they are necessary for any planned agricultural policy and organised food trade, to ensure that as far as is possible the total food requirements of the population are met.

The overall objectives of national and international guidelines for intake of nutrients do not differ from one country or body to another. Recommendations are directed towards an identical aim, which is to ensure, from a nutritional standpoint, the maintenance of health for the majority of the population. Although this common objective is highly commendable in theory, in practice the approaches taken by various countries or bodies may differ significantly. The reasons for this are as follows:

The human population represents a very wide spectrum of physiologic needs. The greater the number of physiologic variants considered, the more heterogeneous the population appears and the smaller is the likelihood that any number of individuals or groups of individuals will share identical nutritional requirements.

To overcome this problem and still ensure the maintenance of health in the majority of the population, it is necessary to subdivide the population into groups and make separate recommendations for each group.

In most countries some 20–30 different population groups are defined, the major variants considered being age, weight, sex, and physical activity, with additional factors introduced for pregnancy. Because of differences in the definition of population groups, recommended daily intakes are not always directly comparable from country to country.

There are also more fundamental reasons for differences in RDI, even within agreed population groups. Firstly, data concerning nutrients have been produced by a variety of different experimental techniques, and secondly, the data that are available are as yet insufficiently detailed to allow for the biological variations that occur within agreed population groups. Thus one group of experts may differ from others in interpreting the data available in the literature, or in its application of safety factors to allow for biological variations. In addition to differences in scientific opinion arising from individual attempts to cover a wide spectrum of physiologic needs, there may also be geographical, agricultural, socio-economic, political, and religious factors which influence RDI tables. It is important to realise that national recommendations are related to the customary diets, availability of foods, and habits of people in the countries. For example, it would be unrealistic to expect the recommendations in a developed Western European country, with a temperate climate and the majority of the population engaged in office work or light mechanised labour, to be the same as those for an underdeveloped country in the tropics, or polar regions, with extremes of climate and the majority of the population engaged in heavy manual labour.

As a result of the above there continue to be differences between various national and international recommendations for dietary intake. These differences are, however, gradually narrowing as more knowledge accumulates and dietary allowances are periodically revised.

It is undoubtedly useful to compare national and international tables of dietary allowances since this will illustrate the areas of disagreement and uncertainty that should be topics for research in the future. For the present, however, it is probable that the use of individual national tables, coupled with dietary surveys, is most likely to help in identifying potential nutritional problems that merit more immediate investigation.

Tables 8 to 15 give the recommended intakes for the following developed countries: United Kingdom, Netherlands, United States, West Germany, Sweden, Canada, South Africa, and Australia. Denmark uses the United States recommendations, and Norway uses Swedish recommendations.

TABLE 8. NETHERLANDS. Recommended calorie and nutrient intakes per person per day, according to Commissie Voedingsnormen, a committee of the Voedingsraad. (Taken from 'Nederlandse Voedingsmiddelentabel'. Voorlichtingsbureau voor de Voeding, 1975.)

Group	Height (cm) [2]	Weight (kg) [2]	Energy (kcal) [3]	Protein (g) [5]	Fat (g)	Carbo-hydrate (g)	Ca (g)	Fe (mg)	A (mg)	β-Carotene (mg)	Thiamin (B₁) (mg)	Riboflavin (B₂) (mg)	Nicotinic acid equiv. [6]	Ascorbic acid (C) (mg)
Children														
0-1 yr incl.	70	9	kg×100	kg×2	kg×5	kg×11	kg×0.1	kg×1	0.15	0.60	kg×0.04	kg×0.06	kg×0.66	30
2-3 yr incl.	93	14	1200	35			0.8	7	0.20	0.80	0.5	0.8	8	35
4-6 yr incl.	114	20	1600	45			0.8	8	0.30	1.10	0.7	1.0	11	35
7-10 yr incl. boys	135	29	2200	60	About 35% fat calories	About 50-55% carbohydrate calories	0.8	10	0.35	1.50	0.9	1.3	15	60
7-9 yr incl. girls	132	27	2000	55			0.8	10	0.35	1.50	0.9	1.2	13	60
Adolescents														
Boys														
11-12 yr incl.	149	38	2600	70			1.2	12	0.45	1.80	1.1	1.6	18	75
13-15 yr incl.	165	51	3000 [4]	80			1.2	15	0.45	2.40	1.2	1.8	20	75
16-19 yr incl.	177	66	3200 [4]	90			1.2	15	0.45	2.40	1.3	2.0	22	75
Girls														
10-12 yr incl.	147	37	2400	65			1.2	12	0.45	1.80	1.0	1.5	16	60
13-15 yr incl.	163	52	2300 [4]	65			1.2	15	0.45	1.80	1.0	1.5	16	75
16-19 yr incl.	166	58	2300 [4]	60			1.0	15	0.45	2.40	1.0	1.5	16	75
Men														
20-35 yr incl.	178	70												
very low activity			2300	65			0.8	10	0.45	2.40	1.0	1.4	16	50
low activity			2600	65	30-35% fat calories	55-60% carbohydrate calories	0.8	10	0.45	2.40	1.1	1.5	18	50
moderate activity			2900	70			0.8	10	0.45	2.40	1.2	1.7	20	50
fairly heavy work			3400	85			0.8	10	0.45	2.40	1.4	2.0	23	50
very heavy work			3800	95			0.8	10	0.45	2.40	1.6	2.3	25	50
35-55 yr incl.	176	70												
very low activity			2100	65			0.8	10	0.45	2.40	0.9	1.3	14	50
low activity			2400	65			0.8	10	0.45	2.40	1.0	1.4	16	50
moderate activity			2700	70			0.8	10	0.45	2.40	1.1	1.6	18	50
fairly heavy work			3200	80			0.8	10	0.45	2.40	1.3	1.9	21	50
55-75 yr incl.	170	68												
very low activity			2000	65			0.8	10	0.45	2.40	0.8	1.2	14	50
low activity			2200	65			0.8	10	0.45	2.40	0.9	1.3	15	50
moderate activity			2500	70			0.8	10	0.45	2.40	1.0	1.5	17	50
fairly heavy work			2900	75			0.8	10	0.45	2.40	1.2	1.7	20	50
75 and older	166	65												
very low activity			1800	65			1.0	10	0.45	1.80	0.8	1.2	12	50
low to moderate activity			2000-2200	65			1.0	10	0.45	1.80	0.9	1.3	14	50

	cm	kg	cal									
Women												
20-30 yr incl.	166	60										
very low activity			1800	55	0.8	12	0.45	2.40	0.8	1.2	12	50
low activity			2000	55	0.8	12	0.45	2.40	0.8	1.3	13	50
moderate activity			2200	60	0.8	12	0.45	2.40	0.9	1.5	15	50
heavy work			2600	70	0.8	12	0.45	2.40	1.1	1.6	17	50
35-55 yr incl.	162	60										
very low activity			1700	55	0.8	12	0.45	2.40	0.7	1.1	11	50
low activity			1900	60	0.8	12	0.45	2.40	0.8	1.2	12	50
moderate activity			2100	65	0.8	12	0.45	2.40	0.9	1.3	14	50
heavy work			2400	65	0.8	12	0.45	2.40	1.0	1.5	16	50
55-75 yr incl.	159	60										
very low activity			1650	55	0.8	12	0.45	2.40	0.7	1.1	11	50
low activity			1850	55	0.8	12	0.45	2.40	0.8	1.2	12	50
moderate activity			2000	55	0.8	12	0.45	2.40	0.9	1.3	13	50
75 yr and older	156	60										
very low activity			1550	55	1.0	10	0.45	1.80	0.7	1.2	10	50
low to moderate activity			1700-1900	55	1.0	10	0.45	1.80	0.8	1.3	12	50
Pregnant women (20-40 yr)												
last 6 months			+100-300	+5-10	+0.5	15	0.55	2.50	1.1	1.6	+2	+25
Nursing mothers exclusive												
breast feeding			+600	+15	+0.7	15	0.65	2.50	1.2	1.8	+5	+25

30-35% fat calories

55-60% carbohydrate calories

1. Protein requirement during the first 6 months depends on the type of feeding and is 2 g of protein/kg of body weight in the case of predominantly milk protein.
2. Figures given are means for the average age group, e.g. 13-15 average = 14 .
3. Because of the wide variations in activity outside school or work, the recommended calorie intake for each group may range from 10% below to 10% above the recommended average.
4. This figure applies for normal physical activity. Agricultural work and training for sports are examples of circumstances requiring increased calorie intake of about 300 calories and 10 g of protein extra. If activity is predominantly low, 300 calories less should be adequate.
5. On the basis of about 11% protein calories for children and adolescents and about 10% protein calories for adults, and assuming adequate protein intake.
6. Figures for Nicotinic acid equivalents are taken from 1971 and 1973 tables as they were not included in the 1975 version.

TABLE 9. UNITED KINGDOM. Recommended daily intakes of energy and nutrients for 1969. (Taken from Department of Health and Social Security Reports on Public Health and Medical Subjects, No. 120, 'Recommended Intakes of Nutrients for the United Kingdom 1969.' Her Majesty's Stationery Office, London.)

Age range [1]	Occupational category	Body weight [3] (kg)	Energy [4] (kcal)	[5] (MJ)	Protein [6] (g)	Thiamin [7] (mg)	Riboflavin (mg)	Nicotinic acid (mg equiv) [8]	Ascorbic acid (mg)	Vitamin A (µg retinol equiv) [9]	Vitamin D [10] (µg cholecalciferol)	Calcium (mg)	Iron (mg)
Boys and Girls													
0–1 year [2]		7.3	800	3.3	20	0.3	0.4	5	15	450	10	600 [12]	6 [12]
1–2 years		11.4	1200	5.0	30	0.5	0.6	7	20	300	10	500	7
2–3 years		13.5	1400	5.9	35	0.6	0.7	8	20	300	10	500	7
3–5 years		16.5	1600	6.7	40	0.6	0.8	9	20	300	10	500	8
5–7 years		20.5	1800	7.5	45	0.7	0.9	10	20	300	2.5	500	8
Boys													
9–12 years		31.9	2500	10.5	63	1.0	1.2	14	25	575	2.5	700	13
12–15 years		45.5	2800	11.7	70	1.1	1.4	16	25	725	2.5	700	14
15–18 years		61.0	3000	12.6	75	1.2	1.7	19	30	750	2.5	600	15
Girls													
9–12 years		33.0	2300	9.6	58	0.9	1.2	13	25	575	2.5	700	13
12–15 years		48.6	2300	9.6	58	0.9	1.4	16	25	725	2.5	700	14
15–18 years		56.1	2300	9.6	58	0.9	1.4	16	30	750	2.5	600	15
Men													
18–35 years	Sedentary	65	2700	11.3	68	1.1	1.7	18	30	750	2.5	500	10
	Moderately active		3000	12.6	75	1.2	1.7	18	30	750	2.5	500	10
	Very active		3600	15.1	90	1.4	1.7	18	30	750	2.5	500	10
35–65 years	Sedentary	65	2600	10.9	65	1.0	1.7	18	30	750	2.5	500	10
	Moderately active		2900	12.1	73	1.2	1.7	18	30	750	2.5	500	10
	Very active		3600	15.1	90	1.4	1.7	18	30	750	2.5	500	10
65–75 years }	Assuming a sedentary life	63	2350	9.8	59	0.9	1.7	18	30	750	2.5	500	10
75 and over }		63	2100	8.8	53	0.8	1.7	18	30	750	2.5	500	10
Women													
18–55 years	Most occupations	55	2200	9.2	55	0.9	1.3	15	30	750	2.5	500	12
	Very active		2500	10.5	63	1.0	1.3	15	30	750	2.5	500	12
55–75 years }	Assuming a sedentary life	53	2050	8.6	51	0.8	1.3	15	30	750	2.5	500	10
75 and over }		53	1900	8.0	48	0.7	1.3	15	30	750	2.5	500	10
Pregnancy, 2nd and 3rd trimester		53	2400	10.0	60	1.0	1.6	18	60	750	10 [11]	1200 [13]	15
Lactation			2700	11.3	68	1.1	1.8	21	60	1200	10	1200	15

1. The ages are from one birthday to another; e.g., 9–12 is from the 9th up to, but not including, the 12th birthday. The figures in the table in general refer to the mid-point of the ranges, though those of the range 18–35 refer to the age 25 years, and for the range 18–55, to 35 years of age.
2. Average figures relating to the first year of life. Energy and minimum protein requirements for the four trimesters are given elsewhere in the DHSS report.
3. The body weights of children and adolescents are averages and relate to London in 1965. The body weights of adults do not represent average values; they are those of the FAO reference man and woman, with a nominal reduction for the elderly.
4. Average requirements relating to groups of individuals.
5. Megajoules (10^6 joules). Calculated from the relation 1 kilocalorie = 4.186 kilojoules, and rounded to one decimal place.
6. Recommended intakes calculated as providing 10 per cent of energy requirements. Minimum protein requirements given in Table 3 of the DHSS report.
7. The figures, calculated from energy requirements and the recommended intake of thiamine of 0.4 mg/1000 kcal, relate to groups of individuals.
8. 1 nicotinic acid equivalent = 1 mg available nicotinic acid or 60 mg tryptophan.
9. 1 retinol equivalent = 1 µg retinol or 6 µg β-carotene or 12 µg other biologically active carotenoids.
10. No dietary source may be necessary for those adequately exposed to sunlight, but the requirement for the housebound may be greater than that recommended.
11. For all three trimesters.
12. These figures apply to infants who are not breast fed. Infants who are entirely breast fed receive smaller quantities; these are adequate since absorption from breast milk is higher.
13. For the third trimester only.

TABLE 10. UNITED STATES OF AMERICA. Recommended daily dietary allowances [1] (revised 1974) according to Food and Nutrition Board, National Research Council. (Taken from "Recommended Dietary Allowances", 8th Edn., 1974.)

	Age (years)	Weight (kg)	Weight (lb)	Height (cm)	Height (in)	Energy (kcal) [2]	Protein (g)	Vitamin A activity (R.E.)[3]	Vitamin A activity (I.U.)	Vitamin D (I.U.)	Vitamin E activity [5] (I.U.)	Ascorbic Acid (mg)	Folacin [6] (µg)	Niacin [7] (mg)	Riboflavin (mg)	Thiamin (mg)	Vitamin B6 (mg)	Vitamin B12 (µg)	Calcium (mg)	Phosphorus (mg)	Iodine (µg)	Iron (mg)	Magnesium (mg)	Zinc (mg)
Infants	0.0–0.5	6	14	60	24	kg × 117	kg × 2.2	420[4]	1400	400	4	35	50	5	0.4	0.3	0.3	0.3	360	240	35	10	60	3
	0.5–1.0	9	20	71	28	kg × 108	kg × 2.0	400	2000	400	5	35	50	8	0.6	0.5	0.4	0.3	540	400	45	15	70	5
Children	1–3	13	28	86	34	1300	23	400	2000	400	7	40	100	9	0.8	0.7	0.6	1.0	800	800	60	15	150	10
	4–6	20	44	110	44	1800	30	500	2500	400	9	40	200	12	1.1	0.9	0.9	1.5	800	800	80	10	200	10
	7–10	30	66	135	54	2400	36	700	3300	400	10	40	300	16	1.2	1.2	1.2	2.0	800	800	110	10	250	10
Males	11–14	44	97	158	63	2800	44	1000	5000	400	12	45	400	18	1.5	1.4	1.6	3.0	1200	1200	130	18	350	15
	15–18	61	134	172	69	3000	54	1000	5000	400	15	45	400	20	1.8	1.5	2.0	3.0	1200	1200	150	18	400	15
	19–22	67	147	172	69	3000	54	1000	5000	400	15	45	400	20	1.8	1.5	2.0	3.0	800	800	140	10	350	15
	23–50	70	154	172	69	2700	56	1000	5000		15	45	400	18	1.6	1.4	2.0	3.0	800	800	130	10	350	15
	51+	70	154	172	69	2400	56	1000	5000		15	45	400	16	1.5	1.2	2.0	3.0	800	800	110	10	350	15
Females	11–14	44	97	155	62	2400	44	800	4000	400	12	45	400	16	1.3	1.2	1.6	3.0	1200	1200	115	18	300	15
	15–18	54	119	162	65	2100	48	800	4000	400	12	45	400	14	1.4	1.1	2.0	3.0	1200	1200	115	18	300	15
	19–22	58	128	162	65	2100	46	800	4000	400	12	45	400	14	1.4	1.1	2.0	3.0	800	800	100	18	300	15
	23–50	58	128	162	65	2000	46	800	4000		12	45	400	13	1.2	1.0	2.0	3.0	800	800	100	18	300	15
	51+	58	128	162	65	1800	46	800	4000		12	45	400	12	1.1	1.0	2.0	3.0	800	800	80	10	300	15
pregnant						+300	+30	1000	5000	400	15	60	800	+2	+0.3	+0.3	2.5	4.0	1200	1200	125	18[8]	450	20
lactating						+500	+20	1200	6000	400	15	80	600	+4	+0.5	+0.3	2.5	4.0	1200	1200	150	18	450	25

1. The allowances are intended to provide for individual variations among most normal persons as they live in the United States under usual environmental stresses. Diets should be based on a variety of common foods in order to provide other nutrients for which human requirements have been less well defined.
2. Kilojoules (kJ) = 4.2 × kcal.
3. Retinol equivalents.
4. Assumed to be all as retinol in milk during the first 6 months of life. All subsequent intakes are assumed to be half as retinol and half as β-carotene when calculated from international units. As retinol equivalents, three-quarters are retinol and one-quarter as β-carotene.
5. Total vitamin E activity, estimated to be 80 per cent as α-tocopherol and 20 per cent other tocopherols.
6. The folacin allowances refer to dietary sources as determined by *Lactobacillus casei* assay. Pure forms of folacin may be effective in doses less than one-quarter of the recommended dietary allowance.
7. Although allowances are expressed as niacin, it is recognised that on average 1 mg of niacin is derived from each 60 mg of dietary tryptophan.
8. This increased requirement cannot be met by ordinary diets; therefore, the use of supplemental iron is recommended.

TABLE 11. WEST GERMANY. Recommended amounts of nutrients per day. Empfehlungen für die Nährstoffzufur. Deutsche Gesellschaft für Ernährung. Umschau Verlag, Frankfurt am Main, 1975.

	Energy [1] (kcal) M	F	(kJ) M	F	Protein (g/kg BW)[4] M	F	Essential fatty acids (g)	Water (ml/kg BW) [4]	Sodium [2] (g)	Chloride [2] (g)	Potassium [2] (mg)	Calcium (mg) M	F	Phosphorus (mg) M	F	Magnesium (mg) M	F	Iron (mg) M	F [3]	Iodine (μg)	Fluoride (mg)
Adults	2600	2200	10900	9200	0.9		10	20–45	2–3	3–5	2–3	800	700	800	700	260	220	12	18	150	1.0
Infants 0–6 mo	600		2500		2.5		2	130–180	0.1–0.3	0.2–0.7	0.3–1.0	500		120–400			75		6	50	0.25
7–12 mo	900		3800		2.2		3	120–145	0.1–0.3	0.2–0.7	0.3–1.0	500		120–400		120			8	50	0.25
Children 1–3 yr	1200		5000		2.2		4	115–125	1–2	2–3	1–2	600		600		130			8	100	0.25–0.5
4–6 yr	1600		6700		2.0		5	100–110	1–2	2–3	1–2	700		700		180			8	100	0.75
7–9 yr	2000		8400		1.8		6	90–100	1–2	2–3	1–2	800		800		220			10	100	0.75
10–12 yr	2400	2100	10000	8800	1.5	1.4	7	70–85	1–2	2–3	1–2	1000	900	1000	900	260	230	12	18	150	1.0
13–14 yr	2700	2400	11300	10000	1.5	1.4	9	50–60	1–2	2–3	1–2	1000	900	1000	900	300	280	12	18	150	1.0
Young people 15–18 yr	3100	2500	13000	10500	1.2	1.0	10	40–50	1–2	2–3	1–2	900	800	900	800	300	250	12	18	150	1.0
Pregnant from 6 mo	2600		10900		1.5		10	20–45	2–3	3–5	2–3	1200	1200	1200		260			25	200	1.0
Lactating	2800		11700		0.9		12	20–45	2–3	3–5	2–3	1200	1200	1200		280			20	200	1.0

	Vitamin A (Retinol-equiv.) (mg) [5]	Vitamin D (μg)	Vitamin E (α-Tocoph.-equiv.) (mg)	Thiamin (mg) [5] M	F	Riboflavin (mg) [5] M	F	Niacin (Niacin-equiv.) (mg)	Vitamin B6 (mg) M	F	Folic Acid (μg)	Pantothenic acid (mg)	Vitamin B12 (μg)	Vitamin C (mg) [8]
Adults	0.9	2.5	12	1.6	1.4	2.0	1.8	9–15	1.8[7]	1.6[7]	400	8	5	75
Infants 0–6 mo	0.6	10	6	0.4		0.5		4	0.3		100	4	0.5	35
7–12 mo	0.7	10	6	0.5		0.6		6	0.5		100	4	1	60
Children 1–3 yr	0.7	5	6	0.7		0.8		8	0.7		200	5	2.5	70
4–6 yr	0.7	2.5	7	1.0		1.1		8	1.1		300	5	2.5	70
7–9 yr	0.8	2.5	8	1.2		1.6		14	1.4		300	6	5	70
10–12 yr	0.8	2.5	10	1.4	1.2	1.9	2.0	14	1.6		400	6	5	75
13–14 yr	0.9	2.5	11	1.4	1.2	1.9	2.0	16	2.1		400	8	5	75
Young people 15–18 yr	0.9	2.5	12	1.6	1.4	2.3	1.9	16	2.1	1.7	400	8	5	75
Pregnant from 6 mo	1.2	10	12	1.6		2.3		12	3.6		800	10	7.5	100
Lactating	2.0	10	20	1.8		2.5		16	2.0		1000	10	7.5	110

1. The values given refer to a 25 year old people with predominantly sedentary occupations.
2. Non-menstruating women: 13 mg of iron.
3. BW = body weight.
4. Approximately 20% losses due to meal preparation have to be considered.
5. Adults over 65 years: 1.1 mg of vitamin A.
6. Adults over 65 years: 2.4 mg of vitamin B6.
7. Approximately 40% losses due to meal preparation have to be considered.

TABLE 12. SWEDEN. Recommendations from the State Institute of Public Health [1]. (Taken from 'Kost och Motion'. Socialstyrelsen, Stockholm, 1971.)

Desirable daily intakes of energy and some nutrients

Age (years) [2]	Reference values Height (cm)	Reference values Weight (kg)	Energy [3] (kcal)	Energy [3] (kJ)	Protein (g)	Calcium (g)	Iron (mg)	Vitamin A Retinol (I.U.)	Vitamin A Retinol (mg)	Thiamin (mg)	Riboflavin (m)	Niacin equiv. (mg) [4]	Ascorbic acid (mg)	Vitamin D (I.U.)	Vitamin D (µg)
Infants															
0–1/6	55	4	kg × 120	kg × 502	kg × 2.2	0.4	6	900	0.27	0.2	0.4	5	35	400	10
1/6–1/2	63	7	kg × 110	kg × 460	kg × 2.0	0.5	10	900	0.27	0.4	0.5	7	35	400	10
1/2–1	72	9	kg × 100	kg × 418	kg × 1.8	0.6	15	900	0.27	0.5	0.6	8	35	400	10
Children															
1–2	81	12	1100	4602	25	0.7	15	1200	0.36	0.6	0.6	8	40	400	10
2–3	91	14	1250	5230	25	0.8	15	1200	0.36	0.6	0.7	8	40	400	10
3–4	100	16	1400	5858	30	0.8	10	1500	0.45	0.7	0.8	9	40	400	10
4–6	110	19	1600	6694	30	0.8	10	1500	0.45	0.8	0.9	11	40	400	10
6–8	121	23	2000	8368	35	0.9	10	2100	0.63	1.0	1.1	13	40	400	10
8–10	131	28	2200	9205	40	1.0	10	2100	0.63	1.1	1.2	15	40	400	10
Men															
10–12	140	35	2500	10460	45	1.2	10	2700	0.81	1.3	1.3	17	40	400	10
12–14	151	43	2700	11297	50	1.4	18	3000	0.90	1.4	1.4	18	45	400	10
14–18	170	59	3000	12552	60	1.4	18	3000	0.90	1.5	1.5	20	55	400	10
18–22	175	67	2800	11715	60	0.8	10	3000	0.90	1.4	1.6	18	60	—	—
22–35	175	70	2800	11715	65	0.8	10	3000	0.90	1.4	1.7	18	60	—	—
35–55	173	70	2600	10878	65	0.8	10	3000	0.90	1.3	1.7	17	60	—	—
55–75+	171	70	2400	10042	65	0.8	10	3000	0.90	1.2	1.7	14	60	—	—
Women															
10–12	142	35	2250	9414	50	1.2	18	2700	0.81	1.1	1.3	15	40	400	10
12–14	154	44	2300	9623	50	1.3	18	3000	0.90	1.2	1.4	15	45	400	10
14–16	157	52	2400	10042	55	1.3	18	3000	0.90	1.2	1.4	16	50	400	10
16–18	160	54	2300	9623	55	1.3	18	3000	0.90	1.2	1.5	15	55	400	10
18–22	163	58	2000	8368	55	0.8	18	3000	0.90	1.0	1.5	13	55	—	—
22–35	163	58	2000	8368	55	0.8	18	3000	0.90	1.0	1.5	13	55	—	—
35–55	160	58	1850	7740	55	0.8	18	3000	0.90	1.0	1.5	13	55	—	—
55–75+	157	58	1700	7113	55	0.8	10	3000	0.90	1.0	1.5	13	55	—	—
Pregnant women			+200	+837	65	+0.4	18	3600	1.08	+0.1	1.8	15	60	400	10
Nursing mothers			+1000	+4184	75	+0.5	18	4800	1.44	+0.5	2.0	20	60	400	10

1. Recommendations are intended to cover individual variation among most healthy people.
2. Recommendations for the age group 22–35 years represent the reference person of 22 years. All others are given for the midpoint of age group.
3. Estimated average energy requirements for persons with modrate physical activity.
4. Niacin equivalents include the vitamin in the diet together with its tryptophan precursor where 60 mg of tryptophan is equivalent to 1 mg of niacin. Animal protein contains about 1.4% tryptophan, vegetable protein about 1% tryptophan.

TABLE 13. CANADA. Dietary standard for Canada. Recommended daily nutrient intake (revised 1975).

Age	Sex	Weight (kg)	Height (cm)	Energy[1] (kcal)	Energy (MJ)[2]	Protein (g)	Thiamin (mg)	Niacin (NE)[6] (mg)	Riboflavin (mg)	Vitamin B6 (mg)	Folate[7] (μg)	Vitamin B12[8] (μg)	Vitamin C (mg)	Vitamin A (RE)[10]	Vitamin D (μg cholecalciferol)[11]	Vitamin E (mg d-α-tocopherol)	Calcium (mg)	Phosphorus (mg)	Magnesium (mg)	Iodine (μg)	Iron (mg)	Zinc (mg)
0–6 mo	Both	6	—	kg×117	kg×0.49	kg×2.2 (2.0)[5]	0.3	5	0.4	0.3	40	0.3	20	400	10	3	500[13]	250[13]	50[13]	35[13]	7[13]	4[13]
7–11 mo	Both	9	—	kg×108	kg×0.45	kg×1.4	0.5	6	0.6	0.4	60	0.3	20	400	10	3	500	400	50	50	7	5
1–3 yr	Both	13	90	1400	5.9	22	0.7	9	0.8	0.8	100	0.9	20	400	10	4	500	500	75	70	8	5
4–6 yr	Both	19	110	1800	7.5	27	0.9	12	1.1	1.3	100	1.5	20	500	5	5	500	500	100	90	9	6
7–9 yr	M	27	129	2200	9.2	33	1.1	14	1.3	1.6	100	1.5	30	700	2.5[12]	6	700	700	150	110	10	7
7–9 yr	F	27	128	2000	8.4	33	1.0	13	1.2	1.4	100	1.5	30	700	2.5[12]	6	700	700	150	100	10	7
10–12 yr	M	36	144	2500	10.5	41	1.2	17	1.5	1.8	100	3.0	30	800	2.5[12]	7	900	900	175	130	11	8
10–12 yr	F	38	145	2300	9.6	40	1.1	15	1.4	1.5	100	3.0	30	800	2.5[12]	7	1000	1000	200	120	11	9
13–15 yr	M	51	162	2800	11.7	52	1.4	19	1.7	2.0	200	3.0	30	1000	2.5[12]	9	1200	1200	250	140	13	10
13–15 yr	F	49	159	2200	9.2	43	1.1	15	1.4	1.5	200	3.0	30	800	2.5[12]	7	800	800	250	110	14	10
16–18 yr	M	64	172	3200	13.4	54	1.6	21	2.0	2.0	200	3.0	30	1000	2.5[12]	10	1000	1000	300	160	14	12
16–18 yr	F	54	161	2100	8.8	43	1.1	14	1.3	1.5	200	3.0	30	800	2.5[12]	6	700	700	250	110	14	11
19–35 yr	M	70	176	3000	12.6	56	1.5	20	1.8	2.0	200	3.0	30	1000	2.5[12]	9	800	800	300	150	10	10
19–35 yr	F	56	161	2100	8.8	41	1.1	14	1.3	1.5	200	3.0	30	800	2.5[12]	6	700	700	250	110	14	9
36–50 yr	M	70	176	2700	11.3	56	1.4	18	1.7	2.0	200	3.0	30	1000	2.5[12]	8	800	800	300	140	10	10
36–50 yr	F	56	161	1900	7.9	41	1.0	13	1.2	1.5	200	3.0	30	800	2.5[12]	6	700	700	250	100	10	9
51+ yr	M	70	176	2300[3]	9.6[3]	56	1.4	18	1.7	2.0	200	3.0	30	1000	2.5[12]	8	800	800	300	140	10	10
51+ yr	F	56	161	1800[3]	7.5[3]	41	1.0	13	1.2	1.5	200	3.0	30	800	2.5[12]	6	700	700	250	100	9	9
Pregnancy				+300[3]	1.3[4]	+20	+0.2	+2	+0.3	+0.5	+50	+1.0	+20	+100	+2.5[12]	1	+500	+500	+25	+15	+1[14]	+3
Lactation				+500	2.1	+24	+0.4	+7	+0.6	+0.6	+50	+0.5	+30	+400	+2.5[12]	+2	+500	+500	+75	+25	+1[14]	+7

1. Recommendations assume characteristic activity pattern for each age group.
2. Megajoules (10⁶ joules). Calculated from the relation 1 kilocalorie = 4.184 kilojoules and rounded to 1 decimal place.
3. Recommended energy intake for age 66+ years reduced to 2000 kcal (8.4 MJ) for men and 1500 kcal (6.3 MJ) for women.
4. Increased energy intake recommended during 2nd and 3rd trimesters. An increase of 100 kcal (418.4 kJ) per day is recommended during the 1st trimester.
5. Recommended protein intake of 2.2 g/kg body wt. for infants age 0.2 mo and 2.0 g/kg body wt. for those age 3–5 mo. Protein recommendation for infants 0–11 mo assumes consumption of breast milk or protein of equivalent quality.
6. 1 NE (niacin equivalent) is equal to 1 mg of niacin or 60 mg of tryptophan.
7. Recommendations are based on estimated average daily protein intake of Canadians.
8. Recommendation given in terms of free folate.
9. Considerably higher levels may be prudent for infants during the first week of life to guard against neonatal tyrosinemia.
10. 1 RE (retinol equivalent) corresponds to a biological activity in humans equal to 1 μg retinol (3.33 I.U.) or 6 μg β-carotene 1 I.U.
11. One μg cholecalciferol is equivalent to 1 μg ergocalciferol (40 I.U. vitamin D activity).
12. Most older children and adults receive vitamin D from irradiation but 2.5 μg daily is recommended. This intake should be increased to 5.0 μg daily during pregnancy and lactation and for those confined indoors or otherwise deprived of sunlight for extended periods.
13. The intake of breast-fed infants may be less than the recommendation but is considered to be adequate.
14. A recommended total intake of 15 mg daily during pregnancy and lactation assumes the presence of adequate stores of iron. If stores are suspected of being inadequate, additional iron as a supplement is recommended.

TABLE 14. SOUTH AFRICA. Recommended minimum daily dietary standards for 1953, according to National Nutrition Council. [Taken from *South African Medical Journal* (1956) 4 Feb, 108–111.]

	Energy (cal) [1]	Protein (g)	Calcium (g)	Iron (mg)	Vitamin A (I.U.)	Thiamin (mg)	Riboflavin (mg)	Niacin (mg)	Ascorbic Adic (mg)
Man (average weight 160 lb)									
Moderately active	3000	65	0.7	9	4000 [11]	1.0	1.6	15	40
Sedentary worker	2300	65	0.7	9	4000 [11]	0.8	1.6	12	40
Heavy worker	4500	65	0.7	9	4000 [11]	1.6	1.6	18	40
Woman (average weight 130 lb)									
Moderately active	2300	55	0.6	12	4000 [11]	0.8	1.4	12	40
Sedentary worker	2000	55	0.6	12	4000 [11]	0.7	1.4	11	40
Heavy worker	2800 [2]	55	0.6	12	4000 [1]	1·0 [2]	1.4	15 [2]	40
Pregnancy, last trimester	[3]								
Moderately active	2600	80 [7]	1.5	15	5000 [12]	0.9	2.0	14	55
Sedentary worker	2200	80 [7]	1.5	15	5000 [12]	0.9	2.0	13	55
Heavy worker	3200	80 [7]	1.5	15	5000 [12]	1.1	2.0	15	55
Lactation	[4]	80 [7]	[9]	15	6000 [12]	[14]	[15]	15	55
Children									
0–3 months	55 cal/lb	1.6 g/lb [8]	0.8 [10]	6 [8]	1500 [13]	0.2	0.5 [8]	2 [16]	20 [17]
4–9 months	50 cal/lb	1.6 g/lb [8]	0.8 [10]	6 [8]	1500 [13]	0.3	0.8 [8]	4	25 [17]
10–12 months	45 cal/lb [5]	1.6 g/lb [8]	0.8 [10]	6 [8]	1500 [13]	0.35	0.9 [8]	4	4 [17]
1–3 years	1100	40	0.6	7	2000 [12]	0.4	1.0	6	40
4–6 years	1500	45	0.7	8	2500 [12]	0.5	1.1	8	40
7–10 years	1900	55	0.8	10	3000 [12]	0.7	1.4	10	40
Girls									
10–12 years	2400 [6]	70	1.0	12	3000 [12]	1.0	1.8	12	40
13–15 years	2600	75	1.2	15	4000 [12]	1.0	1.9	15 [2]	40
16–20 years	2400 [2]	70	1.2	15	4000 [12]	1.0 [2]	1.8	15	40
Boys									
10–12 years	2400	65	0.8	12	3000 [12]	1.0	1.6	12	40
13–15 years	3000	75	1.3	15	4000 [12]	1.2	1.9	15	40
16–20 years	3700	90	1.3	15	4000 [12]	1.5	2.3	15	40

1. Although a single figure is given in each instance, it must be regarded as an average around which there is an individual range. In any case the calorie allowance must be adjusted to the needs of the individual so as to achieve and maintain his desirable weight.
2. In the case of female heavy workers, e.g. those working in fields, this value may have to be appreciably increased.
3. The basal metabolic rate of the mother is not altered by pregnancy and the increased metabolic requirements are due to the foetus and increased weight of the mother.
4. Add 120 calories for each 100 ml of milk produced; e.g. at 1 month the average milk production is about 700 ml and allowance should be made for 840 calories. At 4 months production is about 1 litre and allowance should be made for 1200 calories.
5. For groups under 1 year allow approximately 1000 calories.
6. Allowance has been made for lesser activity compared with boys but this is balanced by greater need because puberty changes set in earlier.
7. An additional allowance of 25 g is made both for pregnancy and full lactation; half of the allowance should be derived from animal sources.
8. The recommendation for infants pertains to protein, iron, etc. derived primarily from cow's milk or commercial milk-preparations.
9. Add 120 mg calcium for each 100 ml of milk produced; e.g. at 4 months the allowance would be 1.2 g.
10. Where breast milk is not being used.
11. Assuming that 1/3rd is present as vitamin A, or 5000 I.U. if 1/5th is present as vitamin A.
12. Preferably 1/3rd in each case as vitamin A.
13. There is evidence that the intake of vitamin A from breast milk, though highly variable, is approximately 2000 I.U. at 2 months and 2500 I.U. at 4 months, but it is not yet established that these intakes are necessary.
14. Allow 0.4 mg for each additional 1000 calories; see Note 4.
15. For full lactation the allowance should be 2.0 mg of riboflavin.
16. This figure is based on the known concentration of niacin in breast milk, and bearing in mind that pellagra amongst children reared entirely at the breast is unknown.
17. Based on the vitamin C concentration in breast milk.

TABLE 15. AUSTRALIA. Dietary allowances for use in Australia (1970 revision) according to National Health and Medical Research Council. (Taken from 'Dietary Allowances for Use in Australia'. Australia Government Publishing Service, Canberra, 1971.)

Subject	Age	Weight (kg)	Weight (lb)	Energy (kcal)	Energy (MJ) [1]	Protein [2] (g)	Calcium (range) (mg)	Iron (mg)	Retinol activity (μg)	Thiamin (mg)	Riboflavin (mg)	Niacin equivalent (mg)	Ascorbic acid ( )	Cholecalciferol Vitamin D (μg)	Vitamin B_{12} [3] (μg)	Folate [3] (μg)
Men	18-35	70	154	2800	11.72	70	400-800	10	750	1.1	1.4	18	30	...	2.0	200
	35-55			2500	10.46					1.0	1.2	16				
	55-75			2100	8.79					0.8	1.0	14				
Women	18-35	58	128	2000	8.37	58	400-800	12	750	0.8	1.0	13	30	...	2.0	200
	35-55			1800	7.53					0.7	0.9	12				
	55-75			1500	6.28					0.6	0.8	10				
Pregnant, 2nd and 3rd trimesters	18-35	+10	+20	2150	9.00	(+8) 66	900-1300	15	750	0.9	1.1	14	60	...	3.0	400
	35+			1950	8.16					0.8	1.0	13				
Lactating	18-35	58	128	2600	10.88	(+20) 78	900-1300	15	1200	1.0	1.3	17	60	...	2.5	300
	35+			2400	10.05					1.0	1.2	16				
Infants	½-1	...	...	110-100 per kg / 50-45 per lb	0.46-0.42 / 0.21-0.19	2.5-0.5 per kg	500-700	4-8	300	0.4	0.5	7	30	10	0.3	60
Children	1-3	13	28	1300	5.44	20-39	400-800	5	250	0.5	0.7	9	30	10	0.9	100
Boys	3-7	19	42	1700	7.12	26-51	400-800	7	350	0.7	0.9	11	30	...	1.5	100
	7-11	28	62	2200	9.21	37-66	600-1100	10	500	0.9	1.1	15	30	...	1.5	100
	11-15	41	90	2900	12.14	51-87	600-1400	12	725	1.2	1.5	19	40	...	2.0	200
	15-18	61	134	3000	12.56	67-90	500-1400	12	750	1.2	1.5	20	50	...	2.0	200
Girls	3-7	18	40	1700	7.12	25-51	400-800	7	350	0.7	0.9	11	30	...	1.5	100
	7-11	27	59	2100	8.79	36-63	600-1100	10	500	0.8	1.1	14	30	...	1.5	100
	11-15	42	92	2500	10.46	52-75	600-1300	12	725	1.0	1.3	17	40	...	2.0	200
	15-18	55	121	2200	9.21	60-66	500-1300	12	750	0.9	1.1	15	50	...	2.0	200

1. 1000 kcal are equivalent to 4.186 million joules or megajoules (MJ).
2. Protein—Adults. For adults in the older age brackets, because of reduced activity and consequently lower kilocalorie requirements, the protein intake and requirements may well be below the selected figure. Adults with comparatively large kilocalorie intakes (e.g. 4000+) within the Australian dietary pattern may have protein intakes appreciably above this figure. It is suggested that, in both of these circumstances, a practical protein allowance be calculated on the basis of 10-12 per cent of the kilocalories being derived from protein.
   Children. The allowance for protein for an individual or a group can be expected to be between these two figures, and because of the character of the Australian dietary pattern, closer to the upper than the lower.
3. Values recommended by the FAO/WHO Expert Group (WHO Technical Report Series No. 452).

4.2 Reasons for the Differences in Recommended Daily Intake

In the following paragraphs some of the factors responsible for the differences in national RDI are illustrated with the aid of the recommendations for protein, vitamin C, and calcium.

Protein

Proteins are large nitrogen-containing molecules, mainly composed of smaller units, the so-called amino acids. Proteins are essential components of all living cells. Examples of high-protein tissues are muscle, liver, skin, and kidneys. Food protein is broken down into its constituent amino acids upon digestion. These amino acids must be constantly supplied to the body tissues, partly for growth of new tissue, partly for maintenance of existing organs. Because the body has a very limited capacity to store reserve proteins, the protein supply must be very regular.

As can be seen from Tables 8–15, recommended daily protein intakes vary with sex and age. Direct comparison of these tables, however, is difficult owing to different ways of presentation. However, if one calculates how many grams of protein per kilogram of body weight are recommended by various countries for male persons of several age groups, as shown in Fig. 20, it is evident that considerable differences exist within each age group. Thirteen-year-old boys in the U.S.A. and Canada need 1.0 g of protein per kg of body weight per day, whereas boys of the same age in the U.K., West Germany and the Netherlands apparently require 1.5 g per kg of body weight daily.

In the making of recommendations for protein intake, two major approaches for establishing minimum protein intakes at which health of nearly all individuals is maintained can be distinguished. The factorial approach, based on numerous detailed experiments on nitrogen balance of humans, seems to be the more acceptable. In this method, protein needed for compensating nitrogen losses via urine, faeces, and skin plus protein needed for growth are calculated on the basis of body weight. In addition, allowances are made for individual variation and for the quality of protein in the local diet. The recommendations for U.S.A., Sweden, and Canada, and the minimum quantities for Australia, are based on such calculations.

The other approach, followed in the Netherlands, United Kingdom, and Australia (for the maximum values), is based on the observation that in the usual Western type of diet between 10 and 13 per cent of the energy is supplied as protein. These levels are thought to be needed for providing sufficiently palatable food. It is questionable whether this level of 10–13 per cent of energy from protein is nutritionally necessary.

For the future, a trend of RDI for protein towards the lower levels given in Fig. 20 seems the reasonable approach, provided that these remain consistent with nitrogen balance under the normal range of living conditions.

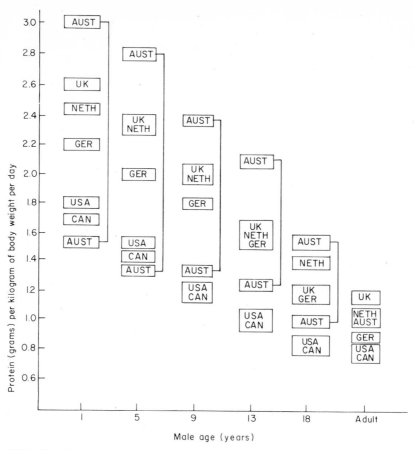

FIG. 20. Protein requirements as recommended in various countries. [The abbreviations in Figs 20–22 refer to Australia, Canada, Germany, Netherlands, South Africa, United Kingdom, and United States of America (=Sweden).]

Vitamin C

Vitamin C is essential to the normal functioning of all cellular units. Only man and other primates, guinea pigs, and some birds are dependent on food sources for the vitamin; other animals and plants synthesise it from glucose. Vitamin C is involved in many aspects of human physiology, some of which are still poorly understood. The actions of the vitamin can be classified in terms of: its reducing capacity, which may contribute to many biochemical reactions; its contributions to the regeneration of tissue, collagen formation, and bone deposition; its participation in the metabolism of folic acid and of aromatic amino acids; its involvement in many other biological processes, such as the level of lipids and cholesterol in the blood, the absorption and transport of iron from blood to tissue, the synthesis of steroids, the activity of leucocytes, and the formation of antibodies.

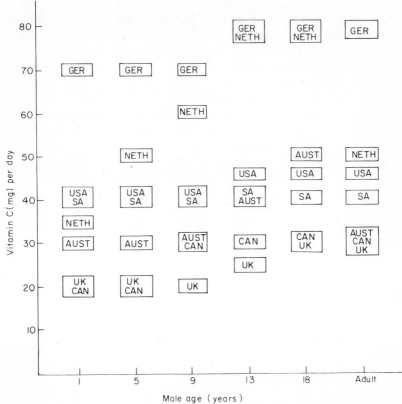

FIG. 21. Vitamin C requirements as recommended in various countries.

In blood plasma the concentration of vitamin C fluctuates with the amount resorbed and may fall rapidly during a period of deficient dietary intake; signs of deficiency (scurvy etc.) however are rather slow to develop. The body is 'saturated' if 1.0–1.5 mg of vitamin C is present per 100 ml of blood plasma; the total amount in the whole body when the blood is saturated, is approximately 5 g.

Recommended daily intake of vitamin C for adults varies from 30 to 75 mg in various countries, as can be seen in Fig. 21. This is due to differences in approach towards the development of an RDI for any nutrient, in various countries, and in the interpretation of the effect of various vitamin C levels. In view of the great number of substances that are reduced by ascorbic acid or oxidised by dehydroascorbic acid, or influenced by the vitamin in another way, it is likely that changes in the concentration of the vitamin may cause changes in e.g. enzyme activity, and may influence the balance in the tissues, and hence the health of the individual.

Specific differences in approach may be illustrated as follows. Many experts agree that the minimal daily intake of vitamin C needed to prevent scurvy is about 10 mg, but this does not mean that the intake then is anywhere near optimal. Higher levels are required to maintain concentration of vitamin C in human tissues characteristic for persons in a given population, living in a given environment and on a given diet with typical quantities of fresh food. The German recommendation is 75 mg for adults, and in the rationale behind this, much emphasis is laid on the losses of vitamin C during preparation and storage of potatoes, vegetables, and fruit, which are the important sources of the vitamin.

The Australian recommendations are based on 'approximate lower levels of the desirable human allowance', with 'the full realisation that the amounts set are several times above the minimal requirements'. The Australian authorities follow their Canadian and British colleagues and recommend 30 mg of vitamin C for adults, bearing in mind 'the lack of evidence to support the hypothesis that blood and tissue saturation with vitamin C represents the optimum state' and many other uncertainties.

The U.S.A. recommendations are based on measurements of the utilisation of vitamin C by healthy men, and provide a generous increment for individual variability and a surplus to compensate for potential losses in food preparation. For adults the RDI is 45 mg. The Dutch recommendations should be seen as 'desirable in terms of the programming of the food supply to a population group'. The figures are 'sufficiently high to match the requirements of almost all individuals in a group, even if their individual requirement is high'.

Efforts to demonstrate the beneficial effects of large doses of vitamin C have not been convincing so far, say the U.S.A. authorities, although the claims are numerous (e.g. prevention of common cold and infections, stimulation of muscular capacity and mental alertness). These claims must therefore await verification and have not yet been taken into account in the RDI in any country. Since vitamin C is multifunctional and at present subject to many studies, it seems that the scientific information will be monitored closely and that the RDI may well be revised if the results of the studies should warrant this.

Calcium

Calcium is a major mineral constituent of the body, making up 1.5–2.0% of body weight in the mature human. The great majority ($>99\%$) of the calcium in the body is present in bones and teeth, while a small proportion is distributed in body fluids and tissues where it participates in the active processes of bone formation and resorption and other important functions such as blood coagulation, neuromuscular irritability, muscle contractibility, and myocardial function.

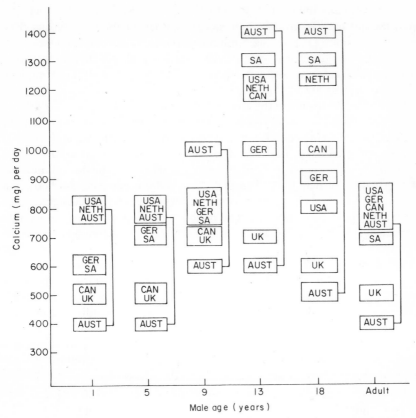

FIG. 22. Calcium requirements as recommended in various countries.

Despite considerable research and the universally recognised importance of calcium in the body and diet, there is often a more than two-fold difference between the RDI for calcium in various countries, as shown in Fig. 22. To understand the reasons for these differences, we need to appreciate some of the complexities of calcium utilisation and the genuine but different scientific approaches that can be taken towards developing an RDI for calcium.

Calcium utilisation and requirement by the body are influenced by many factors, some of which are fairly well identified. Firstly, the age of the individual is obviously important, since with children and adolescents there is a need for calcium retention as bones and teeth are developing, whereas the well nourished adult has no apparent need for more calcium than is required to maintain body stores.

Another important factor is absorption. The calcium content of diet is no direct index of amount actually used by the body, since absorption of calcium from the gut can be strongly influenced by other components in diet. For example, vitamin D, proteins, amino-acids, citric acid, and lactose assist absorption, but oxalic acid, phytic acid, and fatty acids which form insoluble or unavailable calcium salts can depress absorption. Intestinal hypermotility may also depress absorption.

Yet another factor lies in the ability of the human body to adapt to different dietary situations. An adult on a normal mixed diet is usually in a state of calcium equilibrium; i.e. the amount lost in the faeces, urine, and sweat is approximately equal to the amount present in the food. In growing children the body is usually in positive balance with calcium being steadily retained for the formation of new bone. Especially in children, however, the body can adapt; when the body needs more calcium or when the level in diet decreases, a satisfactory balance is maintained through improved efficiency of absorption. Conversely, the body can deal with excess amounts of calcium by a reduction in the efficiency of absorption.

The derivation of RDI for calcium, in the absence of comprehensive data and understanding of the effects of age, growth, other dietary components, body adaptability, etc., is thus a difficult task. To this must be added experimental difficulties. For example, calcium balance studies, which have been a major source of information, have tended to be poor estimates of calcium retention; i.e. retained calcium in such experiments is the relatively small difference between a large intake and a large excretion, and in experiments *in vivo* this situation is inherently inaccurate.

Despite the difficulties, various national and international groups of experts have attempted to derive calcium RDI using a variety of approaches. The FAO/WHO expert group started with an epidemiological approach. It pointed out that most apparently healthy people – children and adults – throughout the world develop, can adapt to, and live satisfactory lives on a dietary intake of calcium that lies somewhere in the wide range 300–1300 mg/day. The group agreed that they could not define a precise minimum but they suggested that a practical allowance was 400–500 mg/day for an adult, and that the rapid bone growth of children called for higher calcium intakes of up to 700 mg/day. The RDI values for calcium in the United Kingdom, which are in the range 500–700 mg/day, are based on similar arguments.

On the other hand, the United States and Australian Research Councils have pointed out that calcium balance studies on adults indicate urinary calcium excretion losses approximating to 175 mg/day, faecel excretion losses approximating to 125 mg/day, and additional losses through the skin as sweat of approximately 20 mg/day. The calcium losses would thus total 320 mg/day. On this basis and assuming 40% absorption, 800 mg would be required to maintain equilibrium in an adult. They therefore suggest that the FAO/WHO figures are a lower limit and that in practice 800 mg/day are required for adults, and amounts up to 1400 mg/day for growing children.

Intermediate between the views of the U.K. and U.S.A. authorities are those for example of the Canadian and South African authorities. The Canadian authorities also started with an epidemiological approach, and considered the adaptation phenomenon where people in many parts of the world could thrive on apparently low calcium intakes. They noted, however, that adaptation to low intakes of calcium was a slow process and that some Canadians would develop calcium deficiency before adapting to severely reduced intakes. The recommended allowance was based, therefore, on the customary calcium level in Canadian diet, with an extra allowance for accelerated skeletal growth at puberty. The Canadian RDI values for calcium thus range from 500 mg/day in infants and 700–800 mg/day for mature adults up to 1200 mg/day for adolescents.

The South African authorities, like their U.S.A. colleagues, have based their data primarily on calcium excretion studies in children and adults. Whilst recognising that the human body adapts itself to levels of calcium considerably lower than the usually recommended allowances, they cite evidence from metabolic studies in well nourished adults that there is a tendency for negative calcium balance at intakes below 10 mg of calcium per kg of body weight per day. They therefore adopt values that approach those in the U.S.A., i.e. 600–1300 mg/day.

All authorities recognise a special situation in pregnancy and lactation. The additional calcium required for the foetus is about 30 g, while 150–300 mg/day is secreted during lactation. This extra requirement is met either by increased efficiency of absorption by the mother or from her bone. To avoid the possible adverse effects of the latter most authorities advocate an additional allowance of up to 700 mg/day during pregnancy and lactation.

The above should illustrate many of the general arguments that have been introduced by various authorities in deriving RDI for calcium. One of the more classic scientific procedures for estimating calcium requirements for infants and children has been a study of skeletal growth from body weight changes. The infant starts life with approximately 30 g of body calcium and the adult body may be assumed to have 1000–1200 g. Assuming a 40% utilisation (as the Australian and U.S.A. authorities do) it would require a daily intake of 350–400 mg to yield 1000–1200 g of body calcium in 20 years. The practical allowances suggested by the FAO/WHO approximate better to this estimate than for example the U.S.A. recommended daily intake.

One further point that may warrant some consideration is the possible significance of dietary calcium in diseases of the bones (e.g. osteoporosis). Loss of bone usually begins at about 45 years of age and may just be an inevitable part of the ageing process. It has been established, however, that there is a high incidence of this disease in the elderly in the U.S.A., particularly amongst women, and it is also common in other countries such as the United Kingdom and Scandinavia. Unfortunately there are no good data on prevalence available, and there are practically no data at all from countries where calcium intake is habitually low. Although high calcium diets are recommended in some countries for the treatment of osteoporosis, there is as yet no agreement amongst experts that calcium therapy is effective against the disease, or that calcium deficiency during growth in early life is related to osteoporosis in later life.

It should be clear from the above that, in the absence of more definitive data on variability in absorption, due to factors such as age, growth, diet, and adaptation, and in the absence of epidemiological evidence which clearly confirms or denies a relationship between calcium intake and longer term disorders such as reduced growth or osteoporosis, disagreements amongst the various national and international groups of experts on the RDI for calcium will remain for the time being. Nevertheless, an appreciation of the basis of these differences can help to point the way to the research required to enable fuller agreement to be met.

4.3 Interpretation of Nutritional Data

The important macro- and micro-nutrients are found in different amounts in different foods. Most foods contain a variety of nutrients but few if any contain them all in sufficient amounts to satisfy the human requirements under normal conditions of food intake.

Table 16 shows for example the average contributions made by groups of foods to the energy value and nutrient content of diet in the United Kingdom. If we now assume quite arbitrarily for illustrative purposes that a significant contribution to diet is more than or equal to 10% of total intake, Table 16 reduces to Table 17, and we see that on average different groups of food play different roles in the supply of our nutrients. For example, as a more general observation, the staple foods meat, milk, bread and other cereals, sugar, and fats are responsible on average for over 70% of our energy and macronutrient intake. In addition these same foods make significant contributions (i.e. 56–72%) to our intake of all but one of the more important vitamins and minerals (the exception is vitamin C). As more specific observations, potatoes, vegetables, and fruit supply on average 87% of the vitamin C intake, and fish, eggs, fats, and dairy products other than milk supply on average 88% of the vitamin D intake.

It is thus not only the amounts but also the variety and nutrient balance of foods that are important in meeting human nutritional requirements. A classic example of this is found in the disease beriberi. This is not necessarily a disease

TABLE 16. Average percentage contributions made by groups of Foods to total nutrient intake in the United Kingdom. (From 'Household Food Consumption and Expenditure,' 1974 Ann. Rept. National Food Survey Committee. H.M.S.O., London, 1976.)

Main food class	Energy	Macronutrients			Vitamins						Minerals	
		Protein	Fat	Carbohydrates	Vitamin A*	Thiamin	Riboflavin	Niacin†	Vitamin C	Vitamin D	Calcium	Iron
Meat	16	29	28	2	23‡	14	19	32	1	1	2	23
Milk	11	18	14	7	13	14	34	12	8	4	47	4
Bread	15	16	2	26	—	22	3	10	—	—	13	19
Other cereals	14	10	8	23	1	22	11	10	—	6	9	14
Sugar and preserves	10	—	—	22	—	—	—	—	2	—	—	1
Fats	15	—	36	—	27	—	—	—	—	43	—	—
Fish	1	4	1	—	—	1	1	4	—	18	1	1
Eggs	2	5	3	—	3	2	8	4	—	17	2	2
Other dairy products	4	7	6	—	7	2	6	4	1	10	15	5
Potatoes	5	4	—	9	—	11	—	9	25	—	1	1
Other vegetables	3	5	1	4	23	8	9	6	29	—	5	9
Fruit	2	1	—	4	1	3	2	1	32	—	2	12
Beverages	—	—	—	1	—	—	5	6	—	1	1	3
Others	2	1	1	2	2	1	2	2	2	—	2	2

*Retinol equivalents.
†Nicotinic acid equivalents.
‡22 from liver.
— <1.

TABLE 17. Food Groups contributing on average 10% or more of total nutrient intake in the United Kingdom. (From 'Household Food Consumption and Expenditure,' 1974 Ann. Rept. National Food Survey Committee. H.M.S.O., London, 1976.)

Main food class	Energy	Macronutrients			Vitamins						Minerals	
		Protein	Fat	Carbohydrates	Vitamin A*	Thiamin	Riboflavin	Niacin†	Vitamin C	Vitamin D	Calcium	Iron
Meat	16	29	28	—	23‡	14	19	32	—	—	—	23
Milk	11	18	14	—	13	14	34	12	—	—	47	—
Bread	15	16	—	26	—	22	—	10	—	—	13	19
Other cereals	14	10	—	23	—	22	11	10	—	—	—	14
Sugar and preserves	10	—	—	22	—	—	—	—	—	—	—	—
Fats	15	—	36	—	27	—	—	—	—	43	—	—
Fish	—	—	—	—	—	—	—	—	—	18	—	—
Eggs	—	—	—	—	—	—	—	—	—	17	—	—
Other dairy products	—	—	—	—	—	—	—	—	—	10	15	—
Potatoes	—	—	—	—	—	11	—	—	25	–	—	—
Other vegetables	—	—	—	—	23	—	—	—	29	—	—	—
Fruit	—	—	—	—	—	—	—	—	32	—	—	—

*Retinol equivalents.
†Nicotinic acid equivalents.
‡22 from liver.
— <1.

of famine; indeed it may occur amongst people with adequate supplies of food and good appetites to eat it. It results in many cases from a poorly balanced diet, in this instance a low thiamin to calorie balance. Though beriberi is atypical of Western Europe, other types of imbalance in diet exist, e.g. the tendency for more affluent people to overeat, and for children to consume excessive amounts of carbohydrate from sweets and other high-sugar desserts.

Assessing the nutritional quality of foods is thus a complex problem. It is not simply a question of commenting that there are x units of a particular vitamin and y units of protein in a product, but also involves consideration of the role of that product in diet (staple food or snack), whether it is replacing an important component within the diet, the type of consumer and the amount usually consumed etc., and the balance of nutrients in the product.

A typical adult daily calorie intake is 2500 calories, and within these total calories should be contained the RDI of each nutrient. Clearly, however, man likes variety in his food, and whilst an overall daily intake, or a major whole meal intake, can be assessed by direct reference to RDI values (or an appropriate fraction for meals), individual items, e.g. meat, peas, ice cream, or fruit pie, cannot be judged by similar direct reference to RDI tables, neither can each be expected to provide the same types of nutrients, since they have different places in diet.

The following sections summarise some of the more important procedures that are required before one can establish the nutritional status of a given product.

(a) Establish the nutrient composition and energy value of the product. This involves the analysis and/or calculation of the contents of those nutrients that can be expected to be present in significant quantities. Although food composition tables are useful in indicating which nutrients are important for a particular product, analysis of the product as consumed is the only reliable approach, especially for products liable to undergo loss of nutrients during processing, storage, and preparation.

(b) Collect data on the daily consumption of the product (average and spread, market penetration, frequency of use for the various consumer groups; such data should come from individual food consumption surveys and from market research.

(c) Calculate the actual daily intake (ADI) of nutrients and energy from the product with the aid of the data mentioned under (a) and (b). Where reliable data on the ADI of a product or group of products are unavailable, one may calculate the intake of nutrients from a single serving of the product. Comparison of the ADI for the product with the ADI from the whole diet illustrates the role of the product in the diet. For example, the role that vegetables play in the nutrient provision in the U.S.A. is shown in Fig. 23.

(d) Compare the ADI of the product, or the nutrient intake per serving for a

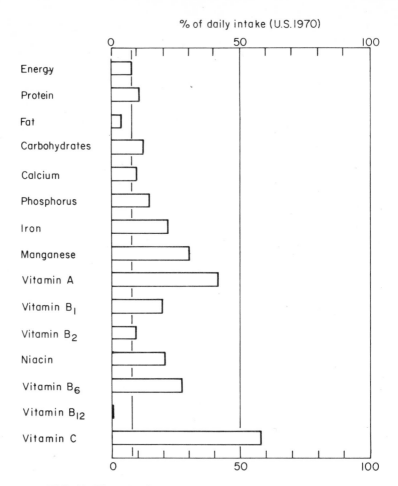

FIG. 23. The role of vegetables in the provision of nutrients.

particular consumer group with the corresponding RDI for this group. The list of ADI values (or nutrient intakes per serving), expressed as percentages of the RDI, which one can call the 'nutrient profile' of the product, is the key to the nutritive value of the product. The percentages for the various nutrients are compared with the percentage for energy intake in determining whether a particular product is a good source of a particular nutrient.

The limited discussion above is intended only as an introduction to this complex area, and to emphasise firstly that expert knowledge and opinion is required, and secondly that each situation must be dealt with in an appropriate way as it arises after careful consideration of a number of aspects.

CHAPTER 5

Food Composition Tables

5.1 The Value and Uses of Food Composition Tables

From a nutritional point of view, it is the amount of nutrients eaten by the consumer that is of paramount importance. The levels of nutrients that finally arrive on the plate are, however, dependent on the total history of the sample. For example, the levels of vitamin C in a serving of cooked cabbage will be dependent to varying degrees upon many of the following factors: the cabbage variety, which relates to plant breeding and genetics; its growth, maturity, and handling at harvest, and its subsequent handling (e.g. post-harvest storage and distribution), which are related to local agricultural practice; its washing, dehydration, freezing, canning, packing, and storage, which are related to factory processing regimes; its distribution and handling in shops; the length and conditions of storage in the home; and finally the method of cooking and the holding time prior to eating, which are related to domestic practice.

The situation is likely to be further complicated by the fact that many of the factors described above may vary from one household to another and from one region to another owing for example to economic and environmental factors and owing to the need to provide foods with the correct attributes to satisfy local consumer attitudes. For example, the degree of sweetness or saltiness, the level of fat, and the extent and method of cooking deemed to be the most desirable often vary from household to household and from region to region.

Bearing in mind the large number of historical variables that may contribute to the levels of nutrients in a food sample, the food analyst should not expect to find that the data in food composition tables will always relate directly to his particular problem.

Nevertheless, food composition tables can give invaluable guidance to the analyst by indicating the range and general levels of nutrients associated with various foods, and the changes to be expected in the levels of nutrients, e.g. during factory processing or home usage.

This in turn will help the analyst to develop a sensible approach to a problem by indicating, for example, which analyses are the most important from a nutritional point of view, the size of sample required, the most appropriate methods to be used, and the concentration range of standards required to form a calibration curve for a particular analysis.

In summary, food composition tables should not be taken literally, but if used with caution they can be most helpful. In the event, however, the food analyst must learn to judge by experience to what extent the enormous volume of data available is likely to be of value and to what extent he needs to generate his own data to solve a particular problem in the most effective manner.

5.2 Sources and Scope of Data

The data in Tables 18–29 represent mean values from sources in the United States of America and a number of Western European countries. It is not intended that these tables should cover all the items of diet in these countries, since this would amount to many thousands of items and add over a hundred pages to this book. Instead, some 10% of all the items of food in the countries reviewed have been included, which represent common and major sources of nutrients in the diet. It is recommended that the food analyst, routinely engaged in nutritional analysis, should equip himself with additional books that deal more exclusively with the subject of food composition (see References section). The values given in Tables 18–29 have been rounded off according to the scheme shown in Fig. 24.

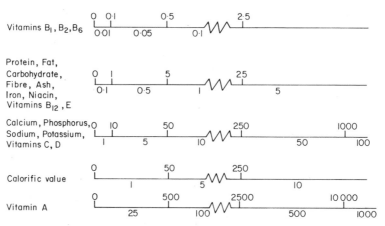

FIG. 24. Rounding-off values for data tables.

5.3. Conversion Factors

1 kcal = 4.184 kilojoules
1 I.U. vitamin A = 0.3 μg of retinol or 0.6 μg of β-carotene
1 I.U. vitamin D = 0.025 μg of vitamin D_2 or D_3

5.4 Abbreviations used in Tables 18–29

CRUD PROT = Crude protein
AVLB CARB = Available carbohydrate
CRUD FIBR = Crude fibre
ENG = Energy
CA = Calcium
P = Total phosphorus
FE = Iron
NA = Sodium
K = Potassium
A = I.U. of vitamin A + β-carotene
THI = Thiamin
RIB = Riboflavin
NI = Sum of nicotinic acid and nicotinamide
ASC ACID = Ascorbic acid and dehydroascorbic acid (vitamin C).
— = Appropriate data not found in present survey
TR = Detected but approaching zero
0 = Not detected.

TABLE 18. Average contents of nutrients and energy per 100 g of food product

MEAT PRODUCTS

COMPONENT	MACRO NUTRIENTS					MINERALS						VITAMINS									
	CRUD PROT G	FAT G	AVLB CARB G	CRUD FIBR G	ENG KCAL	CA MG	P MG	FE MG	NA MG	K MG	ASH G	A I.U.	THI MG	RIB MG	NI MG	B6 MG	B12 µG	ASC ACID MG	D I.U.	E MG	WA TER G
BACON LEAN RAW	13	45	0	0	460	10	140	1.5	1300	240	3.5	TR	0.4	0.15	1.5	0.3	-	0	TR	0.5	38
BACON LEAN FRIED	25	55	0	0	600	10	230	3	2800	500	7	TR	0.4	0.15	1.5	-	-	0	TR	-	12
BACON STREAKY FRIED	25	45	0	0	500	35	230	3	2100	350	6.5	TR	0.45	0.25	3.5	-	-	0	TR	-	21
BEEF CORNED	24	14	0	0	220	15	110	7	1400	120	3.5	TR	0.01	0.2	3.5	0.04	2	0	0	-	58
BEEF LEAN RAW	21	5	0	0	130	10	200	3.5	120	350	1	TR	0.1	0.2	5	0.3	2	0	TR	0.6	73
BEEF LEAN COOKED	28	12	0	0	220	10	220	4.5	70	350	0.8	TR	0.05	0.2	4	0.08	1	0	TR	0.7	59
CHICKEN RAW	21	4	0	0	120	9	200	1.5	70	300	1	50	0.1	0.15	6	0.35	-	0	TR	0.1	73
CHICKEN ROASTED	30	7	0	0	185	15	250	1.5	80	350	1.5	75	0.08	0.15	7	0.15	-	0	TR	0.1	61
HAM COOKED	18	35	0	0	390	10	210	2.5	800	350	2	TR	0.5	0.2	4	0.5	1	0	-	-	45
KIDNEY OX RAW	17	6	0	0	120	10	250	13	220	250	1	800	0.3	2.0	7	0.5	25	15	-	-	75
KIDNEY OX COOKED	25	6	0	0	155	20	300	10	210	240	1	1200	0.5	4.8	11	0.06	20	10	-	-	67
LAMB CHOP RAW	19	9	0	0	155	10	190	2	90	350	1.5	TR	0.15	0.25	5	0.25	-	0	TR	0.8	70
LAMB LEG ROASTED	25	22	0	0	300	7	220	2	70	350	2	TR	0.1	0.25	5	0.1	-	0	TR	-	51
LIVER RAW	19	6	1	0	135	10	350	17	120	350	2	12000	0.4	3.5	15	0.6	40	25	45	1.5	72
LIVER FLOUR DIPPED,FRIED	30	14	3.5	0	260	10	550	25	100	400	2	12000	0.3	4.0	19	0.35	40	15	45	-	50
LUNCHEON MEAT	13	30	0	0	340	20	160	2.5	1000	230	4	TR	0.3	0.2	3	-	-	0	TR	-	49
PORK LEAN RAW	20	7	0	0	145	8	210	2.5	70	350	2.5	TR	0.8	0.2	4.5	0.5	2	0	TR	0.7	70
PORK LEAN ROASTED	24	23	0	0	300	8	250	2.5	70	350	1.5	TR	0.6	0.2	5	0.1	-	0	TR	-	50
SAUSAGE PORK RAW	9	25	10	0	300	10	140	3	750	160	1.5	-	-	-	-	-	-	-	-	-	54
SAUSAGE PORK FRIED	12	23	13	0	300	15	120	1.5	950	230	3	-	0.15	0.1	4	-	-	-	-	-	49
TURKEY RAW	20	15	0	0	215	8	210	3	70	300	1	TR	0.07	0.15	9	0.2	-	-	TR	0.7	64
TURKEY ROASTED	30	13	0	0	235	25	300	3	120	400	1	TR	0.05	0.15	8	1.0	-	-	TR	-	56
VEAL LEAN RAW	20	4.5	0	0	120	10	230	2.5	100	350	1	TR	0.15	0.25	7	0.3	-	0	TR	-	74
VEAL LEAN ROASTED	30	12	0	0	230	15	250	3	90	450	0.8	TR	0.07	0.3	7	-	-	0	TR	-	57

TABLE 19. Average contents of nutrients and energy per 100 g of food product

FISH COMPONENT	MACRO NUTRIENTS CRUD PROT G	FAT G	AVLB CARB G	CRUD FIBR G	ENG KCAL	MINERALS CA MG	P MG	FE MG	NA MG	K MG	ASH G	VITAMINS A I.U.	THI MG	RIB MG	NI MG	B6 MG	B12 µG	ASC ACID MG	D I.U.	E MG	WA TER G
COD STEAMED	18	0.7	0	0	80	10	220	0.5	90	350	1	TR	0.06	0.09	3	0.2	1	1	0	–	80
COD FRIED	25	5	0	0	145	30	250	1	110	400	1.5	TR	0.04	0.1	4	0.2	1	TR	0	–	66
CRAB BOILED	18	3.5	0	0	105	35	250	1	350	250	2	TR	0.15	0.1	2.5	0.35	0.5	TR	0	–	76
EEL SILVER RAW	15	23	0	0	270	15	200	0.8	80	200	1	2500	0.2	0.35	1.5	–	–	TR	1000	–	60
EEL SILVER STEWED	18	30	0	0	340	15	200	0.8	25	200	1.5	4000	0.15	0.35	2	–	1	TR	1500	–	50
HADDOCK RAW	17	0.4	0	0	70	30	210	0.8	90	300	1.5	TR	0.05	0.08	3	0.2	1	TR	0	–	81
HADDOCK FRIED	20	7	4.5	0	160	80	230	1	180	350	3	TR	0.04	0.08	3.5	0.2	1	TR	0	–	66
HADDOCK SMOKED, STEAMED	23	0.7	0	0	100	60	250	1	1200	300	3	TR	0.05	0.07	3.5	0.2	1	TR	0	–	72
HERRING RAW	17	15	0	0	205	100	250	1.5	250	130	2	125	0.02	0.09	3.5	0.45	10	TR	900	–	66
HERRING FRIED	18	14	1.5	0	205	70	250	1.5	90	300	2.5	125	0.03	0.07	3.5	0.2	8	TR	700	–	64
KIPPERS BAKED	23	12	0	0	200	60	350	2	1000	500	4	125	TR	0.2	3.5	–	–	TR	500	–	61
LOBSTER BOILED	20	2.5	0	0	100	60	240	1	250	220	2.5	TR	0.1	0.06	1.5	–	–	TR	0	–	75
MACKEREL FRIED	21	14	0	0	210	15	300	1	150	50	1.5	200	0.1	0.35	8	0.6	4	TR	500	–	63
OYSTERS RAW	8	1	2.5	0	50	120	190	6	350	200	2	300	0.1	0.2	1.5	0.02	15	TR	TR	–	86
PLAICE RAW	16	1.5	0	0	80	15	210	0.8	90	350	1	TR	0.05	0.07	2.5	0.2	1	TR	0	–	81
PLAICE CRUMBED FRIED	18	14	7	0	225	45	250	0.8	120	220	1.5	–	0.04	0.07	2.5	–	1	TR	TR	–	59
PRAWNS	19	2	0	0	95	140	250	1.5	1300	250	4	25	0.04	0.1	2	0.06	–	1	TR	–	75
SALMON PACIFIC STEAMED	19	14	0	0	200	90	350	0.8	80	350	2	250	0.03	0.1	7	0.2	2	TR	500	–	65
SALMON PACIFIC CANNED	20	8	0	0	150	70	300	1.5	550	300	2	250	0.03	0.1	–	0.2	2	TR	500	–	70
SARDINES CANNED	22	22	0	0	290	260	450	3.5	650	450	4	200	0.03	0.2	4.5	0.2	10	TR	300	–	52
SOLE LEMON STEAMED	20	0.9	0	0	90	15	220	0.7	100	300	1.5	TR	0.05	0.07	2.5	0.2	1	TR	300	–	77
TURBOT STEAMED	21	2	0	0	100	15	190	0.5	90	260	1.2	TR	0.02	0.15	3	0.2	2	TR	0	–	75

TABLE 20. Average contents of nutrients and energy per 100 g of food product

FRUIT AND FRUIT JUICES	MACRO NUTRIENTS					MINERALS						VITAMINS									
COMPONENT	CRUD PROT G	FAT G	AVLB CARB G	CRUD FIBR G	ENG KCAL	CA MG	P MG	FE MG	NA MG	K MG	ASH G	A I.U.	THI MG	RIB MG	NI MG	B6 MG	B12 µG	ASC ACID MG	D I.U.	E MG	WATER G
APPLES RAW	0.2	TR	11	1	42	7	9	0.3	2	130	0.3	75	0.03	0.02	0.1	0.05	0	6	0	0.7	85
APPLES STEAMED	0.2	TR	7	1	27	3	15	0.2	3	100	0.3	25	0.03	0.01	0.1	—	—	3	—	—	89
APPLE JUICE	0.1	TR	10	0.1	38	5	7	0.8	3	130	0.2	—	0.01	0.02	0.1	0.01	—	2	0	—	88
APRICOTS RAW	0.9	TR	8	0.6	34	20	20	0.6	1	300	0.7	3000	0.04	0.05	0.7	0.06	0	7	0	—	85
APRICOTS MINCED	0.5	0.1	25	0.4	95	10	15	0.5	4	210	0.4	1700	0.02	0.01	0.4	0.05	0	5	0	—	73
BANANAS	1	0.3	21	0.5	85	8	30	0.5	1	400	0.8	300	0.06	0.06	0.7	0.3	0	10	0	0.4	73
BLACKBERRIES	1	0.3	6	4	29	50	25	0.9	3	190	0.5	190	0.03	0.04	0.4	0.04	0	20	0	—	85
CHERRIES	0.9	0.3	12	0.4	50	20	20	0.4	2	230	0.6	300	0.05	0.06	0.3	0.05	0	8	0	—	76
CRANBERRIES	0.4	0.2	6	1.5	17	15	10	0.8	2	100	0.6	25	0.03	0.02	0.1	—	—	10	0	—	87
CURRANTS RED	1.5	0.2	4.5	3.5	25	30	30	1	3	250	0.6	125	0.06	0.03	0.1	0.06	0	30	0	—	85
CURRANTS BLACK	1.5	0.2	7	2.5	33	60	40	1	3	400	0.9	225	0.04	0.05	0.3	0.08	0	200	0	—	84
DATES	2	0.2	70	2.5	270	60	60	2	5	650	2	50	0.07	0.08	2	0.1	0	2	0	—	20
FIGS	1	0.1	10	1	42	35	25	0.5	3	230	0.7	75	0.06	0.05	0.4	0.15	0	4	0	—	82
GRAPES	0.6	0.2	16	0.5	65	15	25	0.5	2	220	0.4	75	0.04	0.02	0.2	0.09	0	TR	0	—	81
GRAPE JUICE	0.3	TR	18	TR	70	10	10	0.3	2	170	0.4	—	0.04	0.03	0.2	0.03	0	4	0	0.3	81
GRAPE-FRUIT	0.5	0.1	6	0.2	25	15	15	0.3	4	150	0.4	50	0.05	0.02	0.2	0.03	0	40	0	—	90
LEMON	0.4	0.1	3	0.4	15	20	20	0.3	1	140	0.3	TR	0.04	0.01	0.1	0.05	0	45	0	—	89
LEMON JUICE	0.4	0.1	3	0.1	19	7	10	0.2	5	90	0.3	TR	0.02	0.01	0.1	0.06	0	50	0	—	91
MANDARINS IN SYRUP	0.5	TR	17	0.1	65	10	9	0.3	1	90	0.3	100	0.03	0.05	0.2	—	—	3	0	—	81
MELON	0.7	0.1	4	0.3	19	15	20	0.6	15	250	0.5	25	0.04	0.04	0.3	0.05	0	25	0	—	93
OLIVES IN BRINE	1	12	TR	1.5	110	60	15	1.5	2350	70	6	250	TR	0.04	—	0.02	0	—	0	—	77
ORANGES	0.9	0.1	9	0.5	38	40	20	0.3	2	200	0.6	150	0.1	0.04	0.4	0.05	—	50	0	—	86
ORANGE JUICE	0.6	0.1	9	0.3	37	10	20	0.5	1	190	0.5	150	0.09	0.03	0.3	0.03	—	50	0	—	88
PEACHES	0.6	0.1	9	0.6	37	9	25	0.5	3	230	0.5	1100	0.04	0.04	0.8	0.02	—	7	0	—	87
PEACHES IN SYRUP	0.4	0.1	22	0.4	85	3	10	0.3	3	120	0.3	325	0.01	0.02	0.7	—	—	4	0	—	76
PEARS	0.5	0.2	10	1.5	41	10	20	0.2	4	120	0.2	10	0.04	0.03	0.1	0.02	—	4	0	—	84
PEARS IN SYRUP	0.3	0.1	20	0.6	75	5	6	0.4	4	180	0.2	TR	0.01	0.01	0.1	0.01	—	1	0	—	78
PINEAPPLE RAW	0.5	0.1	12	0.4	48	15	8	0.5	1	200	0.4	100	0.08	0.02	0.2	0.09	0	25	0	—	84
PINEAPPLE CANNED	0.3	0.1	20	0.3	75	20	8	0.4	2	80	0.3	50	0.07	0.02	0.2	0.15	—	8	0	—	78
PLUMS RAW	0.5	0.1	10	0.6	40	10	20	0.4	2	170	0.4	250	0.07	0.03	0.5	0.07	0	4	0	—	85
RAISINS	2	0.1	65	0.9	250	60	80	2.5	35	800	2	TR	0.1	0.05	0.5	0.15	0	1	0	—	22
RASPBERRIES	1	0.2	6	3	25	25	25	1	2	160	0.5	125	0.04	0.06	0.5	0.08	0	25	0	—	85
STRAWBERRIES	0.7	0.2	6	1.5	27	20	25	0.7	1	160	0.5	75	0.03	0.06	0.5	0.05	—	60	0	—	90
TANGERINES	0.8	0.1	8	0.5	34	40	20	0.4	2	140	0.4	300	0.06	0.02	0.1	—	—	30	0	—	87

TABLE 21. Average contents of nutrients and energy per 100 g of food product

BEVERAGES

COMPONENT	CRUD PROT G	FAT G	AVLB CARB G	CRUD FIBR G	ENG KCAL	CA MG	P MG	FE MG	NA MG	K MG	ASH G	A I.U.	THI MG	RIB MG	NI MG	B6 MG	B12 µG	ASC ACID MG	D I.U.	E MG	WATER G
ALE (3% ALC)	0.3	TR	3	0	19	8	20	TR	10	30	0.2	TR	TR	0.04	0.6	0.06	-	0	0	-	93
COFFEE INFUSION	0.3	TR	0.4	0	3	8	3	TR	1	100	0.1	0	0	0	0	0	-	0	0	-	99
COLA	0	0	10	0	38	5	20	-	10	-	-	0	0	0	0	0	0	0	0	-	90
LEMONADE	TR	0	6	0	22	5	TR	TR	7	1	0.1	0	0	0	0	0	0	-	-	-	93
ORANGE SQUASH	0.3	TR	35	0.2	130	20	30	0.2	40	70	0.1	0	0	0	0	0	0	-	-	-	61
SHERRY SWEET (15% ALC)	0.3	0	7	0	110	7	10	0.4	15	170	1.5	-	0	0	0	0	-	-	-	-	77
SPIRITS (32% ALC)	TR	0	TR	0	180	0	0	0	0	0	0.2	-	-	-	-	-	-	-	-	-	68
TABLE WINE RED (10% ALC)	0.2	0	0.3	0	60	7	15	0.6	10	150	0.2	0	TR	0.02	0.2	0.05	-	0	0	-	89
TABLE WINE WHITE(10%ALC)	0.1	0	4.5	0	75	10	10	0.7	10	90	0.2	TR	TR	0.01	0.1	0.02	-	0	0	-	85
TEA INFUSION	0.1	0	0	0	1	TR	1	TR	TR	20	TR	0	-	-	-	-	-	0	0	-	99

TABLE 22. Average contents of nutrients and energy per 100 g of food product

MILK, DAIRY PRODUCTS, EGGS

COMPONENT	CRUD PROT G	FAT G	AVLB CARB G	CRUD FIBR G	ENG KCAL	CA MG	P MG	FE MG	NA MG	K MG	ASH G	A I.U.	THI MG	RIB MG	NI MG	B6 MG	B12 µG	ASC ACID MG	D I.U.	E MG	WATER G
MILK WHOLE FRESH	3.5	3.5	5	0	65	120	90	0.1	50	150	0.7	125	0.03	0.15	0.1	0.04	0.3	2	1	0.1	87
MILK FRESH SKIMMED	3.5	0.1	5	0	34	120	90	0.1	50	150	0.7	TR	0.04	0.15	0.1	0.04	0.3	1	TR	TR	90
MILK CONDENSED SWEETENED	8	8	55	0	310	300	220	0.1	100	300	2	350	0.09	0.4	0.2	0.06	0.5	2	4	-	26
MILK DRIED SKIMMED	35	0.5	50	0	330	1300	1000	0.6	550	1500	8	25	0.35	1.5	0.9	0.04	2	9	TR	-	4
BUTTERMILK	3.5	0.2	4	0	31	120	90	TR	90	150	0.7	TR	0.04	0.15	0.1	0.04	-	-	-	-	91
CHEESE CHEDDAR	25	35	TR	0	420	800	500	0.8	650	100	3.5	1300	0.04	0.5	0.1	0.05	2	0	15	1	36
CHEESE DANISH BLUE	22	30	TR	0	360	700	450	0.6	1400	100	4.5	1200	TR	0.45	0.1	-	-	TR	10	1	40
CHEESE EDAM	25	22	1	0	300	700	450	0.4	1100	130	4.5	900	0.03	0.3	0.1	0.06	-	TR	9	-	46
CHEESE CAMEMBERT	20	24	TR	0	300	130	240	0.6	1400	100	4	900	0.04	0.8	1	-	-	0	8	-	50
CHEESE SPREAD	17	22	TR	0	270	500	550	0.6	1400	100	6	900	0.01	0.3	0.1	0.04	-	TR	9	-	51
CREAM SINGLE	3	21	3.5	0	215	90	70	0.3	40	100	0.6	800	0.03	0.15	0.1	0.04	-	1	5	0.4	71
CREAM DOUBLE	2	45	2.5	0	420	60	40	0.2	30	90	0.4	1600	0.02	0.1	0.1	0.03	-	1	15	1	50
YOGHURT LOW FAT	4	2	4.5	0	160	140	100	TR	50	140	0.7	125	0.03	0.2	0.1	0.04	-	1	TR	-	88
EGGS WHOLE RAW	13	12	TR	0	160	60	210	2.5	130	140	1	1000	0.1	0.3	0.1	0.2	0.7	0	70	2	73
EGG WHITE	10	TR	TR	0	40	10	25	0.1	180	140	0.7	0	TR	0.4	0.1	0.01	0.1	0	0	-	88
EGG YOLK	16	30	TR	0	330	140	550	6	50	120	1.5	2800	0.25	0.4	0.1	0.55	2	0	200	-	52
EGG FRIED	14	18	TR	0	220	60	210	2.5	250	160	1.5	1400	0.1	0.3	0.1	-	-	0	-	-	65
EGG POACHED	13	12	TR	0	160	50	210	2.5	200	120	1.5	1200	0.08	0.25	0.1	-	-	0	-	-	73
EGG DRIED POWDER	45	40	3	0	550	200	850	10	450	530	3.5	3600	0.35	1.1	0.1	0.2	-	0	-	-	6

TABLE 23. Average contents of nutrients and energy per 100 g of food product

SOUPS, SAUCES, DRESSINGS, CONDIMENTS

COMPONENT	MACRO NUTRIENTS					MINERALS						VITAMINS									WATER
	CRUD PROT G	FAT G	AVLB CARB G	CRUD FIBR G	ENG KCAL	CA MG	P MG	FE MG	NA MG	K MG	ASH G	A I.U.	THI MG	RIB MG	NI MG	B6 MG	B12 µG	ASC ACID MG	D I.U.	E MG	G
CHEESE SAUCE	7	13	9	0.5	180	200	150	0.2	550	150	0.8	325	0.5	0.3	4	–	–	–	–	–	69
CHICKEN NOODLE SOUP(DRY)	14	8	60	0.4	350	50	150	2.5	5000	150	12	TR	0.01	0.03	0.8	0.02	–	–	–	–	5
CONSOMME	3	1	1	0	25	20	30	0.3	350	50	1	350	0.01	0.04	0.5	0.01	–	–	–	–	93
CREAM STYLE SOUP	2	3.5	7	0.1	65	20	30	0.4	650	130	2.5	–	–	–	–	–	–	–	–	–	84
MUSTARD PREPARED	5	5	4	1	80	80	50	2	600	160	4.5	–	–	–	–	–	–	–	–	–	79
ONION SAUCE	2.5	6	7	0.2	90	80	60	0.2	300	130	0.6	–	–	–	–	–	–	–	–	–	83
PEPPER	9	6	70	11	350	130	30	10	7	40	0.4	225	–	–	–	–	–	–	–	–	3
SALAD CREAM	1	45	10	0	450	10	30	0.2	700	8	2	225	0.01	0.03	TR	–	–	–	–	–	40
SAUCE	2	0.2	24	0.5	100	25	50	1	1900	500	3.5	1400	0.09	0.07	1.5	–	–	5	–	–	67
VINEGAR	0.1	0	0.6	0	3	10	20	0.5	10	90	0.3	–	–	–	–	–	–	–	–	–	94

TABLE 24. Average contents of nutrients and energy per 100 g of food product

VEGETABLES

COMPONENT	MACRO NUTRIENTS					MINERALS						VITAMINS									WATER
	CRUD PROT G	FAT G	AVLB CARB G	CRUD FIBR G	ENG KCAL	CA MG	P MG	FE MG	NA MG	K MG	ASH G	A I.U.	THI MG	RIB MG	NI MG	B6 MG	B12 µG	ASC ACID MG	D I.U.	E MG	G
ASPARAGUS RAW	2.5	0.1	3	0.7	22	20	60	1	4	240	0.6	800	0.15	0.1	0.9	0.05	0	30	0	–	92
ASPARAGUS BOILED DRAINED	2.5	0.1	1	0.7	15	25	70	0.8	2	200	0.4	800	0.15	0.2	1	0.03	–	20	0	–	93
BEANS RUNNER RAW	1.5	0.1	3	1	18	45	35	0.8	7	250	0.7	600	0.2	0.1	0.7	–	–	20	0	–	91
BEANS RUNNER DRAINED	1	0.1	0.9	1	8	40	25	0.6	4	120	0.4	500	0.05	0.08	0.5	–	–	9	0	–	93
BEETROOT RAW	1.5	0.1	7	0.8	33	25	35	0.7	80	350	1	25	0.03	0.04	0.2	0.05	0	7	0	–	88

Food	Water
BEETROOT BOILED	87
BRUSSELS SPROUTS RAW	85
BRUSSELS SPROUTS COOKED	89
CABBAGE SAVOY RAW	91
CABBAGE SAVOY COOKED	95
CARROTS RAW	89
CARROTS BOILED	91
CARROTS CANNED	92
CAULIFLOWER RAW	90
CAULIFLOWER BOILED	94
CELERY RAW	92
CELERY BOILED DRAINED	95
CHICORY RAW	95
CUCUMBER	96
ENDIVES RAW	94
ENDIVES CANNED	96
LEEKS RAW	87
LEEKS BOILED	91
LENTILS RAW	12
LENTILS BOILED	72
LETTUCE	95
MUSHROOMS RAW	90
MUSHROOMS FRIED	64
ONIONS RAW	90
ONIONS COOKED	94
PEAS GREEN FRESH	80
PEAS GREEN COOKED	82
PEAS DRIED RAW	11
PEAS DRIED BOILED	70
PEAS CANNED	70
POTATOES RAW NEW	79
POTATOES RAW OLD	75
POTATOES NEW COOKED	79
POTATOES OLD COOKED	76
POTATOES OLD FRIED	45
SPINACH RAW	92
SPINACH COOKED	88
SPINACH CANNED	92
SWEDES RAW	91
SWEDES BOILED	91
SWEET CORN RAW	71
SWEET CORN BOILED	71
SWEET POTATOES BOILED	71
TOMATOES RAW	94
TOMATOES FRIED	86
TURNIPS RAW	92
TURNIPS BOILED	94

TABLE 25. Average contents of nutrients and energy per 100 g of food product

NUTS

COMPONENT	MACRO NUTRIENTS					MINERALS						VITAMINS									
	CRUD PROT G	FAT G	AVLB CARB G	CRUD FIBR G	ENG KCAL	CA MG	P MG	FE MG	NA MG	K MG	ASH G	A I.U.	THI MG	RIB MG	NI MG	B6 MG	B12 µG	ASC ACID MG	D I.U.	E MG	WA TER G
ALMONDS DRIED	19	55	5	2.5	590	250	450	4.5	5	900	3	0	0.2	0.75	3	0.07	0	TR	0	-	4
BRAZILS	14	65	4.5	3	660	180	650	3.5	4	700	3.5	0	1.0	0.2	0.7	-	-	-	-	-	6
CASHEWS	18	45	15	15	530	45	400	4.5	15	450	2.5	100	0.5	0.2	2	0.35	-	-	-	-	4
CHESTNUTS FRESH	3	2	35	-	160	40	80	2	7	500	1	0	0.35	0.15	0.5	-	-	TR	-	-	53
PEANUTS	25	50	11	2.5	590	60	350	2	5	700	2.5	0	0.4	0.1	16	0.35	0	TR	0	-	4
WALNUTS	14	60	6	2	620	80	400	2.5	9	550	2.5	0	0.35	0.2	2	0.85	0	TR	0	-	5

TABLE 26. Average contents of nutrients and energy per 100 g of food product

OILS AND FATS

COMPONENT	MACRO NUTRIENTS					MINERALS						VITAMINS									
	CRUD PROT G	FAT G	AVLB CARB G	CRUD FIBR G	ENG KCAL	CA MG	P MG	FE MG	NA MG	K MG	ASH G	A I.U.	THI MG	RIB MG	NI MG	B6 MG	B12 µG	ASC ACID MG	D I.U.	E MG	WA TER G
COOKING FAT	0	100	0	0	900	0	0	0	0	0	0	0	0	0	0	0	0	0	0	0	0
BUTTER UNSALTED	0.5	83	0.3	0	750	15	20	0.1	5	20	1.5	3500	TR	TR	TR	-	-	TR	40	2	14
COD LIVER OIL	-	100	-	-	900	-	-	-	-	-	-	76000	-	-	-	-	-	-	-	-	49
MARGARINE SALTED	0.4	83	0.3	-	750	15	15	0.3	500	15	1.5	3000	-	-	-	-	-	-	-	-	14

TABLE 27. Average contents of nutrients and energy per 100 g of food product

SWEETS, JAMS, PRESERVES

COMPONENT	MACRO NUTRIENTS					MINERALS						VITAMINS									
	CRUD PROT G	FAT G	AVLB CARB G	CRUD FIBR G	ENG KCAL	CA MG	P MG	FE MG	NA MG	K MG	ASH G	A I.U.	THI MG	RIB MG	NI MG	B6 MG	B12 µG	ASC ACID MG	D I.U.	E MG	WA TER G
CHOCOLATE MILK	9	35	55	0.4	560	220	220	1.5	160	400	2	175	0.04	0.35	0.6	0.04	-	0	-	-	0
CHOCOLATE DARK	4.5	35	60	0.5	560	90	140	1.5	35	250	1	TR	0.02	0.15	0.3	0.03	-	TR	-	-	0
HONEY COMB	0.4	0	80	0	300	6	20	0.9	7	45	0.3	0	TR	0.04	0.2	0.02	-	2	-	-	19
JAM SEED FRUIT	0.6	0.1	70	1	270	20	15	1.5	15	100	0.3	TR	0.01	0.04	0.1	-	-	6	0	-	28
JAM STONE FRUIT	0.5	0	70	1	260	15	15	1.5	10	100	0.3	TR	0.01	0.03	0.1	-	-	3	0	-	28
JELLY	3	0.1	65	-	260	30	7	1.5	20	50	0.2	TR	0.01	0.02	0.1	-	-	4	0	-	30
MARMALADE	0.4	0.1	65	0.4	245	30	15	0.7	20	60	0.3	75	0.01	0.01	0.1	0.02	-	8	0	-	32
PEANUT BUTTER	30	50	13	2	520	60	400	2	350	700	4	0	0.2	0.09	16	0.45	-	0	-	-	1
SUGAR BROWN	0.2	0	97	0	360	70	20	2	20	220	1.5	-	0.01	0.03	0.2	-	-	0	-	-	1
SUGAR WHITE	0	0	99	0	370	0	0	0.1	1	3	0.1	-	-	-	-	-	-	-	-	-	0
SYRUP	0.3	0	80	0	300	25	20	1.5	250	240	0.7	-	-	-	-	-	-	-	-	-	19
TOFFEES	2	18	75	0	450	90	60	1.5	300	200	1.5	-	-	0.06	0.2	0.2	-	-	-	-	3
TREACLE	1	0	65	0	250	500	30	1.2	1500	1500	4.5	-	0.07	-	-	-	-	-	-	0.2	29

TABLE 28. Average contents of Nutrients and energy per 100 g of food product

CEREAL PRODUCTS COMPONENT	MACRO NUTRIENTS					MINERALS						VITAMINS									WATER
	CRUD PROT G	FAT G	AVLB CARB G	CRUD FIBR G	ENG KCAL	CA MG	P MG	FE MG	NA MG	K MG	ASH G	A I.U.	THI MG	RIB MG	NI MG	B6 MG	B12 µG	ASC ACID MG	D I.U.	E MG	WA TER G
BARLEY PEARL RAW	8	1.5	75	0.5	330	15	200	1.5	4	180	0.9	0	0.09	0.03	3	0.25	-	0	0	-	13
BISCUITS CREAM CRACKERS	8	17	70	0.1	450	115	100	1	500	90	1.5	700	0.03	0.06	0.4	-	-		-	-	3
BISCUITS PLAIN	6	19	65	0.1	440	80	90	0.8	350	110	1.5	700	0.06	0.05	0.4	0.03	-	TR	0	TR	5
BREAD BROWN	9	1.5	50	0.4	240	60	160	2	550	190			0.2	0.1	1.5	0.25	-	0	0	2	36
BREAD WHITE	8	1.5	50	0.2	230	50	90	2	500	110	2		0.15	0.1	2	0.07	-	0	0	-	38
BREAD WHITE TOASTED	10	2.5	60	0.2	290	100	100	2.5	600	120	2	TR	0.2	0.15	2	0.05	-	TR	-	-	24
CAKE PLAIN FRUIT	6	19	55	0.7	400	130	150	1	230	190	1.5	275	0.08	0.1	0.5	0.04	-	TR	-	-	16
CAKE PLAIN SPONGE	8	0.7	55	0	290	50	130	1.5	120	100	0.8	500	0.05	0.1	0.3	0.06	-	TR	-	-	30
CORN FLAKES	8	0.8	85	0.7	340	10	50	1	950	130	2.5		0.15	0.08	2	-	-	0	-	-	7
CORN STARCH	0.4	0.3	85	0	320	7	20	0.7	20	25	0.2						-	0	-	-	12
CUSTARD POWDER	0.3	0.1	85	0.1	340	15	35	0.9	130	35	1.5	TR	0.02	0.01	0.4	TR	-	0	-	-	11
FLOUR ENGLISH	9	2	75	0.6	320	20	200	2	2	230	1.5	175	0.2	0.05	1.5	0.1	0	0	0	-	12
MACARONI	12	1.5	75	0.7	340	25	150	2	10	170	0.7		0.1	0.08	3	0.08	-	0	0	2	10
OAT MEAL	13	8	70	1	390	50	400	4.5	15	350	2		0.5	0.1	1	0.1	0	1	0	-	5
PASTRY FLAKY RAW	5	30	35	0.1	420	50	50	0.1	240	50									-	-	28
PASTRY FLAKY BAKED	7	40	45	0.2	560	70	70	0.1	350	80	1						-	1	-	0	6
PASTRY SHORT RAW	6	30	45	0.1	460	40	50	0.1	450	50	1	0	0.25	0.15	2	-	-	0	0	-	17
PASTRY SHORT BAKED	7	35	50	0.2	530	50	80	0.9	450	60	1	0	0.2	0.15	2	-	-	0	-	0	6
RICE POLISHED RAW	7	0.6	80	0.3	330	15	100	0.5	5	100	0.5	0	0.06	0.03	1.5	0.2	0	0	0	-	11
RICE POLISHED BOILED	2	0.2	25	0.1	105	5	20	0.2	300	30	0.5	0	0.02	0.02	0.4	-	0	0	0	0.4	72
SHORT BREAD	6	25	60	0.2	470	50	100	0.8	150	100	0.6	75	0.04	0.05	0.4	0.03	0	0	-	-	8
SOYA FLOUR(LOW FAT)	45	7	17	2.5	310	250	650	1.5	6	2000	5.3	50	0.85	0.35	2.5	-	0	0	0	-	8
SPAGHETTI DRY	12	1.5	75	0.3	340	25	140	0.4	1	150	0.7	0	0.09	0.06	2	0.1	-	0	-	-	10
SPAGHETTI BOILED	3.5	0.4	23	0.1	105	8	50	0.7	1	60	0.2	0	0.01	0.01	0.3	-	-	0	-	-	72
TAPIOCA	0.5	0.1	85	0.1	320	15	25		4	20	0.2	0	0	0	0	-	-	0	-	-	13
WHEAT GERM	25	9	45	2.5	350	60	1200	8	4	950	4.5	0	2.3	0.8	5	0.6	0	0	-	-	12

TABLE 29. Average contents of nutrients and energy per 100 g of food product

PUDDINGS

COMPONENT	MACRO NUTRIENTS					MINERALS						VITAMINS								
	CRUD PROT G	FAT G	AVLB CARB G	CRUD FIBR G	ENG KCAL	CA MG	P MG	FE MG	NA MG	K MG	ASH G	A I.U.	THI MG	RIB MG	NI MG	B6 MG	B12 µG	ASC ACID MG	D I.U.	WA TER G
APPLE PIE	2	9	35	0.4	220	10	20	0.4	190	90	0.6	25	0.02	0.02	0.4	–	–	1	–	52
BLANC MANGE	3	3.5	19	0	115	120	100	0.2	45	150	0.9	–	–	–	–	–	–	–	–	73
CUSTARD BAKED	5	6	11	0	115	110	120	0.4	80	150	0.8	350	0.04	0.2	0.1	–	–	0	–	76
ICE CREAM	4.5	11	20	0	190	140	110	0.2	70	170	0.9	450	0.04	0.2	0.1	–	–	1	–	62
MILK JELLY	3.5	2	21	0	110	70	50	0.6	35	90	0.5	–	–	–	–	–	–	–	–	73
PANCAKES BAKED	6	10	35	0.1	245	80	120	0.5	260	130	1	125	0.05	0.15	0.4	–	–	–	–	47
RICE PUDDING	3.5	3	25	0.1	135	100	90	0.4	80	180	1	100	0.03	0.15	0.2	–	–	–	–	67
TREACLE TART	3.5	14	60	0.1	370	70	45	0.9	250	160	1.0	–	–	–	–	–	–	–	–	21
TRIFLE	3.5	6	22	0.1	150	80	80	0.9	35	140	0.6	–	–	–	–	–	–	–	–	67

Reference Works for further reading

The following references were used in the preparation of this book, and analysts are advised to consult these for more detailed information on general aspects of nutrition and biochemistry, the analysis of nutrients, recommended intakes of nutrients, and the levels of nutrients in foodstuffs. Users should be aware that new texts on nutrition are being published continually, and should not neglect the current journal literature for the most recent advances. Relevant journals in this respect are (apart from analytical journals) for example:

U.S.A. *Journal of the American Dietetic Association*
  American Journal of Clinical Nutrition
U.K. *Nutrition Reviews*
  British Journal of Nutrition
  Proceedings of the Nutrition Society
Germany *Ernährungsumschau*
Netherlands *Voeding*

General Nutrition

Arlin, M. T. (1972). 'The Science of Nutrition'. Macmillan, New York.
Davidson, S. and Passmore, R. (1975). 'Human Nutrition and Dietetics', 6th Edn. Churchill Livingstone, Edinburgh, London, and New York.
Fischer, P. and Bender, A. (1970). 'The Value of Food'. Oxford University Press, London.
Fleck, H. (1971). 'Introduction to Nutrition', 2nd Edn. Macmillan, London.
Guthrie, H. A. (1975), 'Introductory Nutrition', 3rd Edn. C. V. Mosby Co., St. Louis, Missouri.
Lamb, L. E. (1974). 'Metabolics'. Harper and Row, New York.
Lamb, M. W. and Harden, M. I. (1973). 'The Meaning of Human Nutrition'. Pergamon Press, Elmsford, New York.

Marr, J. S. (1973). 'The Food You Eat'. Evans, New York.

Potter, N. N. (1973). 'Food Science', 2nd Edn. Avi Publ. Co., Westport, Connecticut.

Sebrell, W. H., Jr., Haggerty, J. J. and the editors of *Life* (1967). 'Food and Nutrition'. Time-Life Books, Chicago, Illinois.

Wilson, E. D., Fisher, K. H. and Fuqua, M. E. (1975). 'Principles of Nutrition', 3rd Edn. Wiley, New York.

'Present Knowledge in Nutrition', 3rd Edn. (1967). The Nutrition Foundation, New York.

'The Heinz Handbook of Nutrition' (1965). McGraw-Hill, Maidenhead.

'World Review of Nutrition and Dietetics'. Vols. 1–4 (1959–1963), Pitman Medical Publications, London. Vol. 5 etc. (1965 etc.), Karger, Basel, Switzerland.

Chemistry and Biochemistry

General

Altman, P. L. and Dittmer, D. S. (1968). 'Metabolism'. Federation of American Societies for Experimental Biology, Bethesda, Maryland.

Lehninger, A. L. (1970). 'Biochemistry'. Worth, New York.

Mahler, H. R. and Cordes, E. H. (1971). 'Biological Chemistry', 2nd Edn. Harper, New York and London.

Proteins

Altschul, A. M. (1965). 'Proteins: Their Chemistry and Politics'. Chapman and Hall, London.

Bender, A. E. (1969). Newer methods of assessing protein quality. *Chemistry and Industry*, July.

Gutfrend, H. (1974). 'MIP International Reviews of Science, Biochemistry Series One', Vol. 1 'Chemistry of Macromolecules'. Butterworths, London.

Lawrie, R. A. (1970). 'Proteins as Human Food'. Butterworths, London.

Neurath, H. (1963, 1964). 'The Proteins: Composition, Structure and Function', Vols. 1 and 2. Academic Press, New York.

Schulz, H. W. and Anglemier, A. F. (1964). 'Proteins and Their Reactions'. Symposium on Foods. The Avi Publ. Co., Westport, Connecticut.

Carbohydrates

Aspinall, G. O. (1970). 'Polysaccharides'. Pergamon Press, Oxford.

Aspinall, G. O. (1973). 'MIP International Reviews of Science. Organic Chemistry Series One', Vol. 7 'Carbohydrates'. Butterworths, London.

Birch, G. G. and Green, L. F. (Eds.) (1973). 'Molecular Structure and Function of Food Carbohydrate'. Applied Science Publishers, London.

Coffey, S. (Ed.) (1967). 'The Carbohydrates'. Elsevier, Amsterdam.

Ferrier, R. J. and Collins, P. M. (1972). 'Monosaccharide Chemistry'. Penguin Books, Harmondsworth, London.

Field, A. C. (Ed.) (1973). Fibre in human nutrition. *Proceedings of the Nutrition Society* **32**.

Florkin, M. and Stotz, E. H. (1969). 'Comprehensive Biochemistry', Vol. 17 'Carbohydrate Metabolism'. Elsevier, Amsterdam and New York.

Randle, P. J. (1968). 'Carbohydrate Metabolism'. Academic Press, London.

Rees, D. A. (1967), 'Carbohydrate Polymers'. Oliver and Boyd, Edinburgh.

Spiller, G. and Amen, R. (1975). Dietary fibre in human nutrition. 'CRC Critical Reviews in Food Science and Nutrition', Vol. 7, Issue 1, CRC Press, Oxford.

Whelan, W. J. (1975). 'MIP International Reviews of Science. Biochemistry Series One'. Butterworths, London.

Yudkin, J., Edelman, J. and Hough, L. (1971). 'Sugar: Chemical, Biological and Nutritional Aspects of Sucrose'. Butterworths, London.

Lipids

Ansell, G. B. and Hawthorn, J. N. (1964). 'Chemistry, Metabolism and Function'. Elsevier, Amsterdam.

Aurand, L. W. and Woods, A. E. (1973). 'Food Chemistry', Ch. 5. Avi Publ. Co., Westport, Connecticut.

Dawson, R. M. C. and Rhodes, D. N. (1964). 'Metabolism and Physiological Significance of Lipids'. Wiley, Chichester.

Enselme, J. (1969). 'Unsaturated Fatty Acids and Atherosclerosis'. Pergamon Press, Oxford.

Florkin, M. and Stotz, E. H. (1970). 'Comprehensive Biochemistry', Vol. 18 'Lipid Metabolism'. Elsevier, Amsterdam and New York.

Ganguly, J. and Smellie, R. M. S. (Eds.) (1973). 'Current Trends in the Biochemistry of Lipids'. Academic Press, London.

Gunstone, F. D. (1967). 'An Introduction to the Chemistry and Biochemistry of Fatty Acids and Their Glycerides'. Chapman and Hall, London.

Gurr, M. I. and James, A. T. (1975). 'Lipid Biochemistry: An Introduction', 2nd Edn. Chapman and Hall, London.

Hollingsworth, D. and Russell, M. (Eds.) (1976). 'Nutritional Problems in a Changing World'. British Nutrition Foundation, London.

Johnson, A. R. and Davenport, J. B. (1971). 'Biochemistry and Methodology of Lipids'. Wiley, New York.

Masoro, E. J. (1968). 'Physiological Chemistry of Lipids in Mammals'. Saunders, Philadelphia.

Paoletti, R. and Kritchevsky, D. (Eds.) (1963–1975). 'Advances in Lipid Research', Vols. 1–13. Academic Press, New York, San Francisco, and London.
Vergroesen, A. J. (Ed.) (1975). 'The Role of Fats in Human Nutrition'. Academic Press, London.
Wakil, S. J. (Ed.) (1970). 'Lipid Metabolism'. Academic Press, New York and London.
White, P. L. *et al.* (Eds.) (1975). 'Nutrients in Processed Foods', Vol. 3 'Fats and Carbohydrates'. Publishing Sciences Group, Inc., (a subsidiary of CHC Corporation), Acton, Massachusetts.

Minerals
Mertz, W. and Comatzer, W. E. (Eds.) (1971). 'Newer Trace Elements in Nutrition'. Dekker, New York.
Sanchelli, V. (1969). 'Trace Elements in Agriculture'. Van Nostrand Reinhold, New York.
Underwood, E. J. (1971). 'Trace Elements in Human and Animal Nutrition'. Academic Press, London.
World Health Organisation Technical Report Series No. 532 (1973). 'Trace Elements in Human Nutrition'. W.H.O., Geneva.

Vitamins
Kutsky, R. J. (1973). 'Handbook of Vitamins and Hormones'. Van Nostrand Reinhold, New York.
Sebrell, W. H., Jr. and Harris, R. S. (1967). 'The Vitamins: Chemistry, Physiology, and Pathology Methods', Vols. 6 and 7. Academic Press, New York and London.
Stein, M. (Ed.) (1971). 'Proceedings of the University of Nottingham Residential Seminars on Vitamins'. Churchill Livingstone, Edinburgh and London.

Miscellaneous
'Energy Yielding Components of Foods and Computation of Calorie Values' (1967). Food and Agriculture Organisation of the United Nations.

Food Analysis

General
'Official Methods of Analysis of the Association of Official Agricultural Chemists', 12th Edn. (1975).

Hanson, N. W. (Ed.) (1974). 'Official, Standardised and Recommended Methods of Analysis'. The Society of Analytical Chemistry, London.

Hart, F. L. and Fisher, H. J. (1971). 'Modern Food Analysis'. Springer Verlag, Berlin, Heidelberg, and New York.

Heftmann, E. (Ed.) (1967). 'Chromatography', 2nd Edn. Reinhold, New York.

Joslyn, Maynard A. (Ed.) (1970). 'Methods in Food Analysis', 2nd Edn. Academic Press, New York and London

Lees, R. (1971). 'Laboratory Handbook of Methods of Food Analysis', 2nd Edn. Leonard Hill, London.

Pearson, D. (1970). 'The Chemical Analysis of Foods'. Churchill, London.

Pearson, D. (1973). 'Laboratory Techniques in Food Analysis'. Butterworths, London.

Southgate, D. A. T. (1974). 'Guidelines for the Preparation of Tables of Food Composition'. Karger, Basel, München, Paris, London, New York, and Sydney.

Stahl, E. (Ed.) (1967). 'Thin Layer Chromatography'. Allen and Unwin, London.

Vogel, A. I. (1962). 'A Textbook of Quantitative Inorganic Analysis', 3rd Edn. Longmans, Harlow.

Proteins

Bailey, J. L. (1967). 'Techniques in Protein Chemistry'. Elsevier, Amsterdam, New York, and London.

Blackburn, S. (1970). 'Protein Sequence Determination. Methods and Techniques'. Dekker, New York.

Bradstreet, R. B. (1965). 'The Kjeldahl Method for Organic Nitrogen'. Academic Press, New York and London.

Glick, D. (Ed.) (1966). 'Methods of Biochemical Analysis', Vol. 14 'Determination of Amino Acids'. Wiley-Interscience, London.

Carbohydrates

Goering, H. K. and van Soest, P. J. (1970). 'Forage Fibre Analysis'. Agriculture Handbook No. 379. Agriculture Research Service, U.S. Department of Agriculture.

Korsrud, G. O. and Trick, K. D. (1976). Determination of the available carbohydrate content of carbohydrate reduced and corresponding non-reduced foods. *J. Canadian Inst. Sci. and Technol.* **9**, 92.

Southgate, D. A. T. (1969). Determination of carbohydrates in foods. *J. Sci. Food and Agric.* **20**, 326, 331.

Whistler, R. L. (1962–1963, 1963–1964, 1965, and 1971). 'Methods in Carbohydrate Chemistry', Vols 1–6. Academic Press, New York.

Lipids

Boekenoogen, H. A. (1968). 'Analysis and Characterisation of Oils, Fats, and Fat Products'. Interscience Publishers (a division of John Wiley & Sons Ltd.), London, New York, Sydney.

Christie, W. W. (1973). 'Lipid Analysis'. Pergamon Press, Oxford, New York, Toronto, Sydney, Braunschweig.

Hilditch, T. R. and Williams, P. N. (1964). 'The Chemical Constitution of Natural Fats'. Chapman and Hall, London.

Minerals

Christian, G. D. and Feldman, F. J. (1970). 'Atomic Absorption Spectroscopy: Applications in Agriculture, Biology and Medicine'. Wiley-Interscience, New York and London.

Gorsuch, T. T. (1970). 'The Destruction of Organic Matter'. Pergamon Press, Oxford.

Hubbard, D. P. (Ed.) (1971–1973) and Woodward, C. (Ed.) (1974, 1975). 'Annual Reports on Analytical Atomic Spectroscopy', Vols. 1–5. The Chemical Society, London.

Price, W. J. (1972). 'Analytical Atomic Absorption Spectrometry'. Heyden, London, New York, and Rheine.

Reynolds, R. J., Aldons, K. and Thompson, K. C. (1970). 'Atomic Absorption Spectrophotometry: A Practical Guide'. Griffin, London.

Vitamins

Foud, M. (1966). 'Methods of Vitamin Assay', 3rd Edn., Interscience, Chichester.

McCormick, D. B. and Wright, L. D. (1971). 'Methods in Enzymology', Vol. 18 'Vitamins and Coenzymes'. Academic Press, London.

Strohecker, R. and Henning, H. M. (1963). 'Vitamin Determination'. Verlag Chemie, Weinheim/Bergstr.

Undenfriend, S. (1962). 'Fluorescence Assay in Biology and Medicine'. Plenum Publ., London.

Miscellaneous

Bender, A. E. (1971). The fate of vitamins in food processing operations. 'Proceedings of University of Nottingham Seminars on Vitamins', Session Three. Churchill Livingstone, Edinburgh and London.

Pande, A. (1974). 'Handbook of Moisture Determination and Control'. Dekker, New York.

Recommended Intakes of Nutrients

Department of Health and Social Security Reports on Public Health and Medical Subjects, No. 120 (1969). 'Recommended Intakes of Nutrients for the United Kingdom'.

Food and Nutrition Board, London, National Research Council (1974). 'Recommended Dietary Allowances', 8th Edn.

Empfehlungen Fur Die Nahrstaffyufur. Deutsche Gessellschaft Fur Ernahrung, 1975. Frankfurt am Main: Umschau Verlag.

State Institute of Public Health (1971). 'Kost Och Motion'. Socialstyrelsen, Stockholm.

Dietary Standard for Canada. Recommended Daily Nutrient Intake 1975.

South African Medical Journal (1956). 4 Feb.

National Health and Medical Research Council (1971). 'Dietary Allowances for Use in Australia'. Australia Government Publishing Service, Canberra.

Report of the Committee on International Dietary Allowances of the International Union of Nutritional Sciences (1975). *Nutrition Abstracts and Reviews* **45**, No. 2, 89.

Truswell, A. S. (1976). A comparative look at recommended nutrient intakes. *Proc. Nutr. Soc.* **35**, 1.

Food Composition Tables

Harvey, D. (1970). 'Tables of Amino Acids in Foods and Feedingstuffs', 2nd Edn. Technical Communication No. 19, Commonwealth Bureau of Animal Nutrition.

Jaulones, P. and Harmelli, M. G. (1971). Presence and amounts of nutritional elements in food and the needs of man. *Ann. Nutr. Alim.* **25**, B133–B203.

McCance, R. A. and Widdowson, E. M. (1969). 'Composition of Foods', 2nd Edn. Her Majesty's Stationery Office.

Nobile, S. and Woodhill, J. M. (1973). A survey of the vitamin content of some 2000 foods as they are consumed by selected groups of the Australian population. *Food Technology in Australia*, Feb., p. 80.

Souci, S. W. and Bosch, H. (1967). 'Lebensmittel Tabellen für die Nähwertberechnung'. Wissenschaftliche Vertragegesellschaft, Stuttgart.

Souci, S. W. *et al.* (1969). 'Die Zusammensetzung der Lebensmittel Nähwert Tabellen'. Wissenschaftliche Vertragegesellschaft, Stuttgart.

'Nederlandse Voedingsmiddelentabel' (1975). Voorlichtingsbureau voor de Voeding, The Hague, Netherlands.

'Composition of Foods'. Agriculture Handbook No. 8 (1963). Agricultural Research Service, United States Department of Agriculture.

'Food Composition Tables: Minerals and Vitamins' (1965). Food and Agriculture Organisation of the United Nations.

'Amino Acid Content of Foods and Biological Data on Proteins' (1970). Food and Agriculture Organisation of the United Nations, Rome.

PART II

CHAPTER 6

Methods for the Analysis of Nutrients in Food

CONTENTS

Introduction 103

Section 1   General Sample Preparation 105

Section 2   Moisture and Total Solids

2.1   Air oven method 107
2.2   Vacuum oven method 108
2.3   Dean and Stark distillation 109
2.4   Karl Fischer method 110

Section 3   Proteins and Nitrogenous Compounds

3.1   Total nitrogen and crude protein (Macro Kjeldahl method) 113
3.2   Total nitrogen and crude protein (Automated colorimetric method) 116
3.3   Protein and non-protein nitrogen 120
3.4   Amino acid composition (Automated chromatographic method) 121
3.5   Estimation of protein quality by chemical score (Mitchell and Block method). 128

Section 4   Carbohydrates

4.1   Total available carbohydrate (Manual Clegg Anthrone method) 130
4.2   Total available carbohydrate (Automated Clegg Anthrone method) 131
4.3   Starch and total low molecular weight sugars (Luff–Schoorl method) 134
4.4   Starch (Enzymic method) 139
4.5   Reducing sugars and sucrose (Automated picric acid method) 141
4.6   Lactose (Enzymic method) 144
4.7   Sugar composition (High performance liquid chromatographic method) 146
4.8   Crude fibre 151
4.9   Acid detergent fibre 153

Section 5   Lipids

5.1   Extractable fat (Soxhlet method)   155
5.2   Total fat (Weibul method)   156
5.3   Fat in milk products (Roese–Gottlieb method) .   158
5.4   Total fat (Chloroform–methanol extraction) .   .   161
5.5   Fatty acid composition (Gas–liquid chromatographic method)   163

Section 6   Ash, Elements and Inorganic Constituents

6.1   Ash   166
6.2   Wet digestion of food products for element analysis   167
6.3   Metals (Atomic absorption method) .   .   .   168
6.4   Potassium and sodium (Flame photometric method)   171
6.5   Chloride (Rapid Volhard method)   173
6.6   Chloride (Potentiometric method)   174
6.7   Ammonia (Colorimetric method)   175
6.8   Total phosphorus (Colorimetric method) .   .   178
6.9   Total phosphorus (Colorimetric method, automated)   180

Section 7   Fat-Soluble Vitamins

7.1   Vitamin A and/or β-carotene (Manual method) .   183
7.2   Vitamin A and/or β-carotene (Automated high performance liquid chromatographic method) .   188
7.3   Vitamin D (Gas chromatographic method) .   .   191
7.4   Vitamin E (Thin layer chromatographic method) .   197

Section 8   Water-Soluble Vitamins

8.1   Vitamin B_1 (Manual method)   201
8.2   Vitamin B_2 (Manual method)   205
8.3   Vitamin B_1 (Thiamin) (Automated high performance liquid chromatographic method)   208
8.4   Vitamin B_2 (Automated high performance liquid chromatographic method)   210
8.5   Niacin (Automated high performance liquid chromatographic method)   213
8.6   Introduction to microbiological assay techniques for the determination of vitamins   216
8.7   Vitamin B_2 (Riboflavin) (Microbiological method).   218
8.8   Niacin (Microbiological method)   221
8.9   Vitamin B_6 (Microbiological method) .   .   .   224
8.10 Vitamin B_{12} (Microbiological method) .   .   .   227
8.11 Ascorbic acid (Visual titration method) .   .   .   230
8.12 Ascorbic acid (Electrometric method) .   .   .   232
8.13 Ascorbic and dehydroascorbic acid (Fluorimetric method)   235

Section 9   Calculation of Calorific Value   239

Introduction

To be effective, the analyst needs to be familiar with both the principles behind and the practice of the methods he uses. The methods that follow are therefore presented in a way which allow him to do this most readily. Firstly, each method contains either a short statement of the principles behind the more well-known procedures (e.g. the determination of protein, using the Kjeldahl method) or more detailed background information for less well-known procedures (e.g. high performance liquid chromatography methods for vitamins). Secondly, full details are given of practice of the method in terms of apparatus and reagents required and the stepwise procedures needed to perform the analysis correctly which are derived from experience in the authors' laboratories.

As far as possible methods have been chosen that are suitable for applications to a wide range of foodstuffs. However, one cannot anticipate all situations that may arise and the analyst may have to use his experience to modify and add further methods where appropriate.

General Sample Preparation

Application
   The method is applicable to all types of foodstuffs and ingredients.

Principle
   Samples of foodstuffs are made homogeneous prior to analysis by a procedure that protects labile nutrients. The sample is first frozen and then broken into small pieces. After further cooling with liquid nitrogen the sample is ground to a fine powder.
Note. If vitamin C analyses are required they must be done directly after the homogenisation procedure.

Reagents
   Liquid nitrogen.

Apparatus
   Sample bottles. 200 ml, wide mouth.
   Deep freeze. At $-20°C$.
   Meat mincer.
   Waring blender. About 12 000 rotations/minute.
   Stainless steel beakers for the Waring blender. Volume 1 litre.
   Dewar vessels. Volume 2 litres.
   Plastic containers. Styrofoam, large enough to contain several sample bottles.

Procedure
  1. Remove bones or other materials that are not generally consumed.
  2. Freeze the sample as quickly as possible at a temperature of $-20°C$. Store the sample at $-20°C$ until homogenisation.
  3. Break the frozen sample into small pieces (e.g. with a wooden mallet). Keep fozen through complete procedure.

4. Take a representative sample of at least 500 g.
5. Fill the Dewar vessel with liquid nitrogen.
6. Pass sample through mincer (screw and housing cooled to about $-20°C$) and immediately transfer into the Dewar vessel.
7. Cool the beaker of the Waring blender with liquid nitrogen.
8. Homogenise the sample until a homogeneous powder is obtained (Notes 1 and 2).
9. Cool the wide-mouthed sample bottles to $-20°C$.
10. Transfer the powder into the pre-cooled sample bottles.
11. Store the bottle in a plastic container and store the whole in a freezer at $-20°C$ (Note 3) until required for analysis.

Notes
1. Take care—nitrogen gas escapes from the beaker during this procedure.
2. If necessary prepare separate samples for metal, vitamin, and other analyses.
3. Storage in a plastic container prevents great temperature fluctuations which give rise to evaporation of water from the sample onto the wall of the glass bottle.

Moisture and Total Solids

2.1 Air Oven Method

Application

The method is applicable to all food products except those that may contain volatile compounds other than water or those liable to decompose at 100°C.

Principle

The sample is dried to constant weight in an air oven.

Apparatus

Oven. Temperature 100–102°C.

Dishes. Nickel, stainless steel, aluminium, or porcelain. Metal dishes should not be used when the substance to be dried may have a corrosive action. Whenever possible, dishes with well-fitting lids should be used.

Desiccator. Containing dry phosphorus pentoxide, calcium chloride, or granular silica gel.

Procedure

1. Dry the empty dish and lid in the oven for 15 minutes and transfer to the desiccator to cool (about 10 minutes for metal dishes; about 20 minutes for porcelain dishes). Weigh the empty dish and lid to the nearest mg.
2. Mix the prepared sample thoroughly and transfer about 5 g to the dish. Replace the lid and weigh the dish and contents, as rapidly as possible, to the nearest mg.
3. Remove the lid and place the dish and lid in the oven, avoiding contact of the dish with the walls. Dry for 6 hours. For products that do not decompose during long periods of drying, it is permissible to dry overnight, that is about 16 hours.

4. Remove the dish from the oven, replace the lid, cool in a desiccator, and re-weigh when cold.
5. Dry for a further hour to ensure that constant weight has been achieved.

Calculation

$$
\begin{aligned}
\text{Let: Weight (g) of sample} &= W_1 \\
\text{Loss of weight (g)} &= W_2 \\
\text{Weight (g) of dried sample} &= W_3 \\
\text{Then: Moisture (\%)} &= (W_2/W_1) \times 100 \\
\text{Total solids (\%)} &= (W_3/W_1) \times 100
\end{aligned}
$$

2.2 Vacuum Oven Method

Application

The method is applicable to food products that contain compounds liable to decompose at 100°C.

Principle

Many products that decompose in an oven at 100°C can be dried at a lower temperature under reduced pressure. The efficiency of the method depends on maintaining as low a pressure as possible in the oven and on removing the water vapour from the oven quickly.

Apparatus

Vacuum oven. Thermostatically controlled and connected through a drying train to a vacuum pump capable of maintaining the pressure in the oven below 25 mm of mercury. The oven should be provided with an air-inlet connected to a sulphuric acid gas-drying bottle and trap for releasing the vacuum.

Dishes. Metal dishes with close-fitting lids and flat bottoms to provide maximum area of contact with the heating plate.

Desiccator. Containing dry phosphorus pentoxide, calcium chloride, or granular silica gel.

Procedure

1. Dry, cool, and weigh a dish and lid to the nearest mg.
2. Transfer to the dish and spread out about 5 g of well mixed sample. Replace the lid and weigh the dish and contents as rapidly as possible, to the nearest mg.
3. Partially uncover the dish and place in the oven, evacuate the oven, and dry the sample at 70°C for 6 hours. If the material contains an appreciable amount of water, it is an advantage to bleed a slow flow of dry air into the oven for the first hour or two.

4. After the period of heating, admit a slow stream of dry air to the oven until atmospheric pressure is reached, cover the dish, transfer to the desiccator to cool, and re-weigh when cold.
5. Dry for a further hour to ensure that constant weight has been achieved.

Calculation

$$
\begin{aligned}
\text{Let: Weight (g) of sample} &= W_1 \\
\text{Loss of weight (g)} &= W_2 \\
\text{Weight (g) of dried sample} &= W_3 \\
\text{Then: Moisture (\%)} &= (W_2/W_1) \times 100 \\
\text{Total solids (\%)} &= (W_3/W_1) \times 100
\end{aligned}
$$

2.3 Dean and Stark Distillation

Application
  The method is applicable to fatty foods and those containing significant volatiles other than water.

Principle
  Water, along with xylene or toluene, is distilled over in a Dean and Stark apparatus at a constant boiling temperature.

Reagents
  Xylene or toluene.

Apparatus
  Dean and Stark distillation apparatus (see Fig. 1 and Note 1).
  Heating mantle.

Procedure
1. Weigh 10–20 g (Note 2) and place in a 250 ml flask.
2. Add 100 ml of xylene or toluene to the flask.
3. Connect the flask to the condenser and Dean and Stark apparatus.
4. Pass cooling water through the condenser, heat the flask and contents on the heating mantle, and bring the contents to the boil.
5. Adjust the heating mantle so that the flask contents are just kept boiling, and continue heating for at least $1\frac{1}{2}$ hours.
6. Switch off the heating mantle and allow the apparatus to cool, especially the side arm.
7. Record the volume of water in the side-arm. (Note 3).

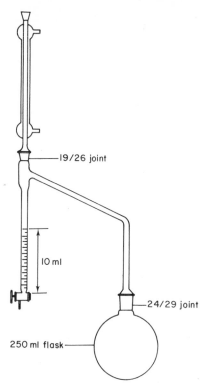

FIG. 1. Dean and Stark distillation apparatus.

Calculation

Let: Weight (g) of sample taken $= W$
           Volume (ml) of water collected $= V$
Then: Moisture (%) $= (V/W) \times 100$

Notes

1. The apparatus should be kept clean and free from grease.
2. The amounts of water in the sample must be less than 10 ml.
3. Some droplets of water may collect above the graduated arm at the bottom of the condenser. These should be dislodged by a wire or thin glass rod before measuring the volume.

2.4 Karl Fischer Method

Application

Determination of moisture content in dehydrated products.

Principle

The water in the sample is titrated by means of the Karl Fischer reagent which consists of sulphur dioxide, pyridine, and iodine in anhydrous methanol. The reagent is standardised against the water of crystallisation of hydrated sodium acetate. The end-point of the titration is detected electrometrically, utilising the 'dead stop' end-point technique.

Reagents

Methanol. Anhydrous for Karl Fischer titration, contains 1% pyridine (Note 1).

Sodium acetate trihydrate.

Karl Fischer reagent. From laboratory suppliers (Note 2). The reagent is standardised daily by the use of hydrated sodium acetate in place of the sample, by the process described below.

Apparatus (Note 3)

Burette. All glass, automatic filling type, fully protected against moisture ingress.

Electrometric apparatus and galvanometer. Suitable for 'dead stop' end-point technique.

Titration vessel. Any convenient vessel that can be provided with agitation either by injection of dry inert gas or by magnetic stirrer. All excess of air must be excluded by maintaining a small positive pressure of inert gas (nitrogen or carbon dioxide).

Procedure

1. Weigh, to the nearest mg, an amount of sample containing approximately 100 mg water, into a pre-dried round-bottomed 50 ml flask.
2. Pipette 40 ml of methanol into the flask, quickly place it on the heating range and connect the reflux condenser (the condenser should be 'conditioned' before use by refluxing methanol in a flask for 15 minutes and then allow to drain for a further 15 minutes).
3. Gently boil the contents of the flask under reflux for 15 minutes.
4. Remove the flask from the heat and, with the condenser still attached, allow to drain for 15 minutes.
5. Remove and stopper the flask.
6. Pipette a 10 ml aliquot of the extract into the titration vessel and titrate (Note 4) with Karl Fischer reagent (titrate duplicate aliquots where possible) to the 'dead stop' end-point and record volume of titrant used.
7. Determine a blank titre by taking a 10 ml aliquot from 40 ml methanol which has been refluxed as described above (steps 2 to 5).

Standardisation of Karl Fischer Reagent

The water content ($M\%$) of hydrated sodium acetate ($CH_3COONa.3H_2O$) is accurately determined by oven drying at 120°C for 4 hours.

1. Weigh, to the nearest mg, approximately 0.4 g of hydrated sodium acetate into a pre-dried 50 ml round-bottomed flask.
2. Pipette 40 ml of methanol into the flask, and stopper immediately.
3. Swirl the contents until the added sample is completely dissolved.
4. Titrate 10 ml aliquots of this solution, and 10 ml of methanol blank as described in the procedure (steps 6 and 7).

Calculation

(*a*) *Water equivalent* (*F*) *of the Karl Fischer reagent*:

Water content (%) of sodium acetate	$= M$
Weight (g) of sodium acetate trihydrate	$= W$
Volume (ml) of titrant used for standard	$= V_s$
Volume (ml) of titrant used for blank	$= V_b$

Then: The water equivalent of Karl Fischer reagent F
(mg water per ml) $= (W \times M \times 2.5)/(V_s - V_b)$

(*b*) *Determination of water content in sample*:

Let: Weight (g) of sample	$= W_1$
Volume (ml) of reagent used for sample	$= V_1$
Volume (ml) of reagent used for blank	$= V_2$
Standardisation factor of reagent (mg water per ml)	$= F$

Then: Water content (%) of sample $= [0.4 \times F \times (V_1 - V_2)]/W_1$

Notes

1. If it is necessary to dry methanol for the preparation of the Karl Fischer reagent or for use as solvent, do so by distillation from a small quantity of magnesium and a few crystals of iodine.
2. Although the component parts of the apparatus are detailed, it is convenient to buy the complete set-up from laboratory suppliers if Karl Fischer titrations are to be done regularly.
3. All titrations should be made quickly, with as little delay as possible between additions of reagent.

SECTION 3

Proteins and Nitrogenous Compounds

3.1 Total Nitrogen and Crude Protein (Macro Kjeldahl Method)

Application
   The method is applicable to all food products.

Principle
   The product is digested with concentrated sulphuric acid, using copper sulphate as a catalyst, to convert organic nitrogen to ammonium ions. Alkali is added and the liberated ammonia distilled into an excess of boric acid solution. The distillate is titrated with hydrochloric acid to determine the ammonia absorbed in the boric acid.

Reagents
   Sulphuric acid. Concentrated, sp gr. 1.84. Nitrogen-free.
   Hydrochloric acid, 0.1N, standardised.
   Boric acid solution. Dissolve 40 g of boric acid (H_3BO_3) in water and dilute to 1000 ml.
   Sodium hydroxide solution. Carbonate free, containing approximately 33 g of sodium hydroxide per 100 g of solution. Prepare by dissolving 500 g of sodium hydroxide in 1000 ml of water.
   Copper sulphate pentahydrate.
   Potassium sulphate, anhydrous.
   Mixed indicator solution. Dissolve 2 g of methyl red and 1 g of methylene blue in 1000 ml of ethanol (96% v/v). The colour change of this indicator solution occurs at a pH of 5.4. Store the indicator solution in a brown bottle in a dark and cool place.
   Boiling regulators. For the digestion: glass beads, silicon carbide or splinters of hard porcelain. For the distillation: silicon carbide or freshly ignited pieces of pumice stone.
   Grease-proof paper. Pieces about 9×6 cm.

Apparatus

Kjeldahl flask. Capacity about 800 ml provided, if desired, with a pear-shaped glass bulb loosely fitting into the neck of the flask.

Distillation apparatus. Steam or direct.

Heating device. Such that the Kjeldahl flask can be heated in an inclined position and that the source of heat only touches that part of the flask wall which is below the liquid level. For gas heating a plate of asbestos provided with a circular hole, so that only the lower part of the flask is exposed to the free flame, is suitable.

Procedure

(*a*) *Digestion*:
1. Place a few boiling regulators in the Kjeldahl flask and add 15 g of potassium sulphate and 0.5 g of copper sulphate.
2. Weigh, to the nearest mg, about 2 g of the prepared sample (or 1.5 g of a sample rich in fat) on a piece of grease-proof paper.
3. Transfer the paper and the test portion to the Kjeldahl flask.
4. Add 25 ml of sulphuric acid and mix by gently swirling the liquid. If desired, a pear-shaped glass bulb may be inserted into the neck of the flask with its tapered end downwards.
5. Place the flask on the heating device at an angle of about 40° from the vertical.
6. Heat the flask gently until foaming has ceased and the contents have become completely liquefied.
7. Digest by boiling vigorously, occasionally rotating the flask, until the liquid has become completely clear and of a light blue-green colour (see Note 1).
8. Keep the liquid boiling for another 1.5 hours. The total digestion time should not be less than 2 hours. Take care that no condensed liquid runs down the outside of the flask.
9. Cool to about 40°C and add cautiously about 50 ml of water, mix, and allow to cool. Proceed to either instruction b.1 or instruction c.1.

(*b*) *Steam distillation*:
1. Add 50 ml of the boric acid solution to a 500 ml conical flask, add 4 drops of the indicator solution, mix and place the flask under the condenser of the distillation apparatus so that the outlet of the adapter dips into the liquid.
2. Transfer the contents of the Kjeldahl flask to the distillation flask and rinse the Kjeldahl flask with about 50 ml of water.

3. Pour carefully, down the inclined neck of the distillation flask, 100 ml of sodium hydroxide solution from a measuring cylinder so that the two layers in the flask do not mix. Immediately attach the flask to the splash head of the distillation apparatus.
4. Pass steam through the alkaline liquid, at first slowly to reduce foaming, until it boils. Keep the liquid boiling for 20 minutes. At least 150 ml of distillate should be collected (Note 2).
5. Lower the conical flask just before terminating the distillation, so that the outlet of the adapter is above the liquid level.
6. Rinse the outlet of the adapter above the liquid (internally and externally) with a little water; verify the completion of the ammonia distillation by testing the distillate from the condenser with red litmus paper.
7. Stop heating (Note 3).

(c) *Direct distillation*:
1. As instruction b.1.
2. Dilute the contents of the Kjeldahl flask cautiously with about 300 ml of water and swirl. If desired, transfer to a 1 litre flask. Add fresh boiling regulators.
3. After about 15 minutes add 100 ml of sodium hydroxide solution from a measuring cylinder carefully down the inclined neck of the flask so that the two layers in the flask do not mix. Immediately attach the flask to the splash head of the distillation apparatus.
4. Distil at least 150 and a maximum of 250 ml of distillate. If the mixture bumps irregularly after 150 ml of distillate have been collected discontinue the distillation. (Note 2).
5. Lower the conical flask just before terminating the distillation, so that the outlet of the adapter is above the liquid level.
6. Rinse the outlet of the adapter above the liquid (internally and externally) with a little water; verify the completion of the ammonia distillation by testing the distillate from the condenser with red litmus paper.
7. Stop heating (Note 3).

(d) *Titration*:
1. Titrate the contents of the conical flask with the hydrochloric acid solution.
2. Record the volume, to the nearest 0.02 ml, of hydrochloric acid required.

(e) *Blank test*:
1. Conduct a blank test in duplicate following the procedure except for addition of the sample (Note 4).

Calculation

Let: Weight (g) of the test portion $= W$
      Volume (ml) of hydrochloric acid solution
      required for the blank test $= V_1$
      Volume (ml) of hydrochloric acid solution
      required for the test portion $= V_2$
      Normality of hydrochloric acid $= N$

Then: Total nitrogen (%) $= \dfrac{(V_2 - V_1) \times N}{W} \times 1.4$

      Crude protein (%) $= \dfrac{(V_2 - V_1) \times N}{W} \times 1.4 \times 6.25$

where 6.25 is the general factor (Note 5).

Notes
1. Avoid overheating during the digestion since this will result in an excessive loss of sulphuric acid and a low recovery of nitrogen.
2. Make sure that the distillate is cooled effectively and that the boric acid solution does not become warm.
3. If the distillation is found to be incomplete, carry out a new determination following the instructions carefully.
4. Blank tests should be run when fresh batches of reagents or freshly prepared solutions are used and occasionally when reagents and solutions have been in use for some time.
5. The crude protein value is calculated from total nitrogen using the appropriate factor: general 6.25; meat products 6.25; milk products 6.38; cereal products 5.70.

3.2 Total Nitrogen and Crude Protein (Automated Colorimetric Method)

Application
The method is applicable to all types of food products.

Principle
The product is digested with concentrated sulphuric acid, with hydrogen peroxide and mercury as catalyst, to convert organic nitrogen into ammonium ions. Using an automated system, phenol and sodium hypochlorite react with the ammonia to give a blue colour which is measured at 510 and 630 nm. By measurements at two wavelengths it is possible to cover a factor of 100 in nitrogen content.

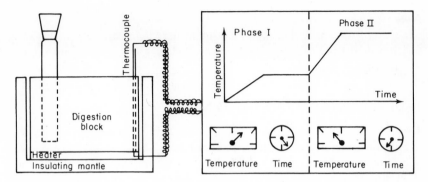

FIG. 2. Unit for Kjeldahl digestion in nitrogen determination.

A. DIGESTION

Reagents

Kjeltabs. Supplier Thompson and Capper, Liverpool. Composition: 5 g of potassium sulphate and 0.35 g of mercury oxide.

Sulphuric acid, concentrated (sp. gr. 1.84).

Hydrogen peroxide, 30% w/w.

Apparatus

Digestion block. As in Fig. 2; see Procedure for conditions.

Tubes, for the digestion blocks: capacity 100 or 250 ml, graduated.

Procedure

1a. Weigh 1 g of sample containing no more than 30 mg of crude protein nitrogen and transfer quantitatively to a 100 ml tube, or

1b. Weigh 1 g of sample containing more than 30 mg of crude protein nitrogen and transfer quantitatively to a 250 ml tube.

2. Add 1 Kjeltab to the tube, followed by 2 ml of hydrogen peroxide and 10 ml of sulphuric acid.

3. Mix thoroughly.

4. Place the tubes in the digestion block.

5. Heat at a mean temperature of 130°C (Note 1) for 30 minutes.

6. Increase the heat over a 30 minute period until the digest is boiling (about 350°C).

7. When the digest is colourless, continue boiling for a further 30 minutes.

8. Remove the tubes from the block and cool to room temperature.

9. Dilute to the mark and cool again to room temperature.

10. Mix thoroughly.

11. Check dilution to the mark. Adjust if necessary.

12. Prepare a blank using the same amounts of reagents.

Note •
1. Care. Products with a high fat content may foam strongly at this stage. Lower the pre-digestion temperature below 130°C and extend the time beyond 30 minutes.

B. DETERMINATION OF AMMONIUM

Reagents and standards

Sulphuric acid, concentrated (sp. gr. 1.84).

Sodium hydroxide, 1N. Dissolve 40 g of sodium hydroxide in water and dilute to 1000 ml.

Sodium hydroxide–EDTA solution. Dissolve 40 g of sodium hydroxide and 6 g of ethylenediaminetetraacetic acid (EDTA) in 800 ml of water and dilute to 1000 ml.

Sodium hydroxide–phenol solution. Weigh 50 g of phenol in a 400 ml flask. Dissolve 20 g of sodium hydroxide in 250 ml of water. Add the sodium hydroxide solution to the phenol. Dissolve the phenol and transfer the solution quantitatively to a 1 litre graduated flask; add 200 ml of ethanol, mix, and dilute to the mark with water.

Sodium nitroprusside solution. Weigh 200 mg of sodium nitroprusside, dissolve in water, and dilute to 1000 ml.

Alcohol, 96% w/v.

Standard ammonium sulphate solution. Weigh 4.714 g of ammonium sulphate $[(NH_4)_2SO_4]$, transfer to a 1 litre graduated flask, and dilute to the mark.

Working standard solutions. Prepare solutions containing 1, 5, 10, and 35 mg of N per 100 ml in 10% sulphuric acid. Pipette 1, 5, 10, and 35 ml aliquots respectively of standard solution into 100 ml graduated flasks. Dilute all flasks to approximately 35 ml with water. Carefully add 10 ml of concentrated sulphuric acid (sp. gr. 1.84) to each flask, keeping the flask cool, and dilute to volume with water. Mix.

Apparatus

Continuous flow analyser for automatic analysis with appropriate accessories.

Procedure

1. Assemble the apparatus as in Fig. 3. Pump the reagents through until the system is equilibrated.
2. Fill the sampler cups with working standard solutions followed by sample solutions. The range of standards should be repeated after every 12–15 samples.

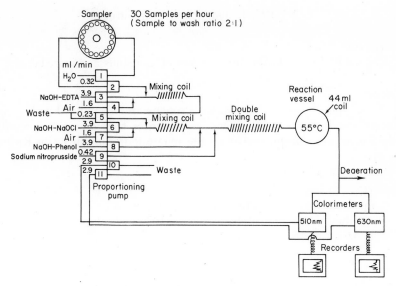

FIG. 3. Set-up for automatic nitrogen determination.

3. Run the system until all samples have passed the flow-cell of the colorimeters.

Calculation

(*a*) *Calibration curve*:
   Record the absorbance values of blanks and calibration solutions. Correct the absorbances of standards and samples for the blank value. Prepare two calibration curves, one of the low range (630 nm) the other for the high (510 nm).

(*b*) *Total nitrogen and crude protein content*:
   Read the nitrogen values of the samples from the calibration curve.

Let: This value $= a$
       Weight (g) of sample $= W$
Then: If 100 ml tubes are used:
       Total nitrogen content (%) $= a/(10 \times W)$
       Crude protein content (%) $= [a/(10 \times W)] \times 6.25$
      If 250 ml tubes are used:
       Total nitrogen content (%) $= a/(4 \times W)$
       Crude protein content (%) $= a/(4 \times W) \times 6.25$
       See Note 5 of Method 3.1 for protein conversion factors.

E

3.3 Protein and Non-protein Nitrogen

Application
This method is applicable for all types of foodstuffs and ingredients.

Principle
The sample is digested with water. Proteins are precipitated with copper acetate and non-protein nitrogenous substances remain in solution. After filtration nitrogen is estimated on the precipitate and/or filtrate by the Kjeldahl method.

Reagents
Copper acetate monohydrate, 3% w/v solution.
Aluminium potassium sulphate $24H_2O$, 10% w/v solution.
Silicone anti-foam.

Apparatus
Kjeldahl flasks, 800 ml.
Filter funnels, diameter about 4 inches or 12 cm.
Buchner flasks.
Filter paper. Whatman No. 541 or Schleicher and Schüll No. 1505, diameter about 18 cm.

Procedure
 1. Weigh or pipette an amount of sample according to one of the following instructions: 2 g for up to 25% crude protein; 1 g for 25–50% crude protein; 0.5 g for above 50% crude protein. Milks are gently shaken, warmed in a water bath to 40°C, gently shaken again, and cooled to 20°C; then a suitable aliquot is pipetted out, usually 11 ml.
 2. Transfer the sample to the Kjeldahl flask.
 3. Add approximately 50 ml distilled water, a few anti-bumping granules, and one or two drops of silicone antifoam.
 4. Digest the mixture by boiling gently for half an hour (take care not to boil it dry).
 5. While the digest is still hot add 2 ml of aluminium potassium sulphate solution, and swirl to mix.
 6. Re-heat to just boiling.
 7. Add 50 ml of copper acetate solution, and mix thoroughly.
 8. Allow to cool.
 9. Filter the contents through a filter paper using a 4 inch funnel and Buchner flask. Wash the flask and precipitate with 50 ml of cold water.
10. For protein nitrogen return the paper and residue to the original flask and

determine the nitrogen content by the Kjeldahl method. Beware of excess frothing at the initial digestion stage (see Method 3.1).

11. For non-protein nitrogen transfer the filtrate from the Buchner flask to a clear Kjeldahl flask and determine the nitrogen content by the Kjeldahl method (see Method 3.1 or 3.2).

Calculation

The results of the Kjeldahl analyses give the protein and non-protein nitrogen contents.

3.4 Amino acid Composition (Automated Chromatographic Method)

Application

The amino acid composition of proteins in foodstuffs.

Principle

(a) *Treatment of the proteins to release the amino acids*:

Samples are hydrolysed with hydrochloric acid, in the absence of air, to break the peptide bonds of a protein. This procedure gives good results for the acid-stable amino acids, i.e. all those commonly occurring in food proteins except cystine, cysteine, methionine, and tryptophan, which are labile under acid hydrolysis conditions, and require separate methods of analysis.

The amino acids cystine, cysteine, and methionine are therefore first oxidised with performic acid, under controlled conditions, to convert into their residues of cysteic acid and methionine sulphone. These acid-stable residues are then freed from the protein by hydrolysis with hydrochloric acid.

Hydrolysis of the sample with barium hydroxide in solution in the absence of air releases tryptophan without decomposition.

(b) *Chromatography and determination of the amino acids*:

The chromatographic analyses are performed on an automatic amino acid analyser (BioCal BC 200; Note 1) which is based on the principles described by Spackman *et al.* (1958) whereby amino acids are fractionated on ion-exchange resins, by elution with a discrete buffer system. The eluate is mixed with ninhydrin reagent, and the mixture is passed through the reaction coil immersed in a boiling water bath. After a 15 minute delay, during which the colour reaction takes place, the emerging stream passes through the colorimeter, where the absorbence is monitored at 570 and 440 nm. A schematic diagram is shown in Fig. 4.

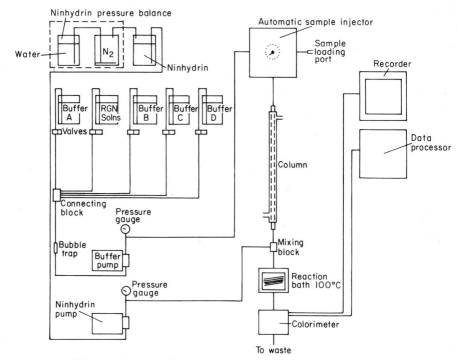

FIG. 4. Schematic diagram of an automatic amino acid analyser.

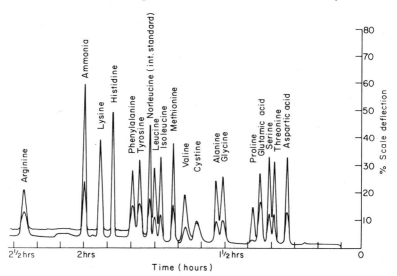

FIG. 5. Chromatogram of a standard mixture containing 0–0.4 μmole of each amino acid.

The areas of the recorded peaks are measured either by the use of the data processor or by triangulation.

The amount of amino acid in a sample is determined by comparison of areas of unknown sample solutions with areas obtained from a standard solution containing known amounts of amino acids (Fig. 5).

A. HYDROLYSIS AND OXIDATION

Reagents
1. *General*
   Hydrochloric acid, concentrated (sp. gr. 1.18).
   Hydrochloric acid, 6N. Dilute 531 ml of concentrated hydrochloric acid (sp. gr 1.18) with water.
   Hydrochloric acid, 0.1N. Dilute 10 ml of 6N hydrochloric acid to 600 ml with water.
   Sample buffer, pH 2.20. Dissolve the following, and make up to 1 litre with water: 19.6 g of sodium citrate dihydrate; 20.0 ml of thiodiglycol; 2.0 ml of Brij 35 (30% w/v); 16.2 ml of concentrated hydrochloric acid (sp. gr. 1.18); and 0.1 ml of caprylic acid. Adjust to pH 2.20 with concentrated hydrochloric acid.
   Filter paper. Whatman No. 542 or Schleicher and Schüll No. 1507.
2. *For sulphur amino acid analysis*
   Hydrobromic acid, 48%.
   Performic acid. Mix together 18 ml of formic acid and 2 ml of hydrogen peroxide (30%). Allow to stand for 1 hour, then cool to 0°C.
   Hydrochloric acid, 0.05N. Dilute 5 ml of 6N solution to 600 ml with water.
   Sample buffer, pH 2.81. Dissolve the following and make up to 1 litre with water: 10.5 g of citric acid monohydrate; 12.6 g of lithium hydroxide; 10.0 ml of thiodiglycol; 2.0 ml of Brij 35 (30% w/v); and 20.0 ml of concentrated hydrochloric acid (sp. gr. 1.18). Adjust to pH 2.81 with concentrated hydrochloric acid.
3. *For tryptophan analysis*
   Barium hydroxide. $8H_2O$.

Standards
   General standard. The standard solution is prepared in pH 2.20 sample buffer, to contain 0.04 μmol per ml of each amino acid.
   Sulphur amino acid standard. The standard solution is prepared by dissolving both cystine and methionine in the minimum amount of 6N hydrochloric acid, and then diluting with water to give a solution to contain 0.4 μmol per ml of each.
   Tryptophan standard. The standard solution is prepared in pH 3.25 buffer (buffer E, section b) solution to contain 0.1 μmol per ml of tryptophan.

Apparatus
  1. *General*
     Electrothermal heating rack.
     Rotary evaporator.
     pH meter.
  2. *For sulphur amino acid analysis*
     Ice bath, temp. 0°C.
  3. *For tryptophan analysis*
     McCartney bottles.
     Oven or autoclave, temp. 120°C.

Procedure

(*a*) *Acid-stable amino acids*:
  1. Weigh a quantity of sample accurately into a 100 ml round-bottom, side-arm flask.
     (a) if greater than 20% crude protein, 35 mg sample.
     (b) if less than 20% crude protein, 70 mg sample.
  2. Add 70 ml of 6N hydrochloric acid, connect a nitrogen bleed tube into the side arm, and reflux for 20 hours under a water condenser.
  3. Wash down the condenser with water, disconnect the flask, and allow to cool.
  4. Filter the hydrolysate through a glass-wool plug, into a 100 ml volumetric flask, and with water washings make up to volume.
  5. Pipette an aliquot of hydrolysate, containing approximately 2.5 mg of crude protein, into an appropriate flask, and rotary evaporate to dryness at 45°C.
  6. Add 10 ml of water, and re-evaporate to dryness.
  7. Dissolve the residue in pH 2.20 sample buffer, transfer to a volumetric flask, and with washings make up to volume.
  8. Filter the solution through a filter paper.

(*b*) *Sulphur amino acids*:
  1. Weigh a quantity of the sample accurately (approx. 100mg).
  2. Pipette 5 ml of sulphur amino acid standard.
  3. Transfer sample and standard to 100 ml round-bottom wide-neck flasks. Add 20 ml of performic acid to both sample and standard, and leave to stand in an ice bath at 0°C:
     (a) for soluble proteins leave for 4 hours.
     (b) for insoluble proteins leave overnight.
  4. Whilst still at 0°C add 2.5 ml of 48% hydrobromic acid, to destroy excess performic acid, and swirl gently.

5. Immediately rotary evaporate to dryness at 40°C; 100 ml of 1N sodium hydroxide solution should be present in the condenser flask to absorb bromine, which distils over.
6. Add about 5 ml of 0.05N hydrochloric acid; re-evaporate to dryness.
7. Add about 5 ml of water; re-evaporate to dryness.
8. Add, to the flask containing the oxidised sample, 70 ml of 6N hydrochloric acid solution, connect a nitrogen bleed tube, and reflux for 20 hours under a water condenser.
9. Follow the steps (a)3–(a)6 inclusive.
10. Dissolve the residue in pH 2.81 sample buffer, and transfer to a 50 ml volumetric flask; with washings make up to volume.
11. Filter the solution through a filter paper.

(c) *Tryptophan*:
1. Weigh a quantity of sample, to contain approximately 4 mg of tryptophan, into a McCartney bottle.
2. Weigh into the same bottle 8.4 g of barium hydroxide.
3. Add 16 ml of boiled water.
4. Heat to 100°C under a stream of nitrogen, to remove air, and stopper the bottle.
5. Heat at 120°C for 8 hours.
6. Cool, and acidify with concentrated hydrochloric acid to pH 3–4.
7. Dilute to 50 ml.
8. Filter the solution through a filter paper.

B. CHROMATOGRAPHIC ANALYSIS

Reagents

Buffer A, pH 3.25, sodium ion concentration 0.2N. Dissolve the following amounts of salts in water: 19.6 g of sodium citrate dihydrate; 5.0 ml of thiodiglycol; 2.0 ml of Brij 35 (30% w/v); 40.0 ml of ethanol; 12.0 ml of conc. hydrochloric acid (sp. gr. 1.18); and 0.1 ml of caprylic acid. Make up to 1 litre with water and adjust pH to 3.25 with concentrated hydrochloric acid (Note 2).

Buffer B, pH 4.10, sodium ion concentration 0.2N. Dissolve the following amounts of salts in water: 19.6 g of sodium citrate dihydrate; 2.0 ml of Brij 35 (30% w/v); 8.4 ml of concentrated hydrochloric acid (sp. gr. 1.18); and 0.1 ml of caprylic acid. Make up to 1 litre with water and adjust pH to 4.10 with concentrated hydrochloric acid.

Buffer C, pH 6.55, sodium ion concentration 1.2N. Dissolve the following amounts of salts in water: 19.6 g of sodium citrate dihydrate; 58.5 g of sodium chloride; 2.0 ml of Brij 35 (30% w/v); and 0.1 ml caprylic acid. Make up to 1 litre with water and adjust pH to 6.55 with concentrated hydrochloric acid.

Buffer D, pH 2.81, lithium ion concentration 0.3N. Dissolve the following amounts of salts in water: 10.5 g of citric acid monohydrate; 12.6 g of lithium hydroxide monohydrate; 2.5 ml of thiodiglycol; 2.0 ml of Brij 35 (30% w/v); and 20.0 ml of conc. hydrochloric acid (sp. gr. 1.18). Make up to 1 litre with water and ajust pH to 2.81 with concentrated hydrochloric acid.

Buffer E, pH 3.25, sodium ion concentration 0.2N. Dissolve the following amounts of salts in water: 19.6 g of sodium citrate dihydrate; 5.0 ml of thiodiglycol; 2.0 ml of Brij 35 (30% w/v); 12.0 ml of conc. hydrochloric acid (sp. gr. 1.18); and 0.1 ml of caprylic acid. Make up to 1 litre with water and adjust pH to 3.25 with concentrated hydrochloric acid.

Sodium hydroxide solution, 0.4N. Dissolve 8.00 g of sodium hydroxide and make up to 500 ml with water. (Regeneration, RGN, solution)

Lithium hydroxide solution, 0.3N. Dissolve 3.50 g of lithium hydroxide monohydrate and make up to 250 ml with water. (Regeneration, RGN, solution)

Ninhydrin colour reagent. The following reagents are mixed in a conical flask, by addition in the order shown: (i) 1500 ml of methyl cellosolve; (ii) 500 ml of pH 5.6 acetate buffer (4N sodium acetate adjusted to pH 5.6 with acetic acid); (iii) 40 g of ninhydrin; and (iv) 0.8 g of stannous chloride. During the mixing process, nitrogen should be continuously bubbled through the solution. A period of approximately 10 minutes should be left between each addition to enable thorough mixing and deaeration. Continue the nitrogen flow through the mixed reagent, to avoid contact with air, until the solution is placed in the storage reservoir. To give better baseline stability at high sensitivities, the reagent should be filtered through a sintered glass funnel (porosity 4), before transfer to the storage reservoir.

Bio-Rad A5 and Bio-Rad A6.

Sephadex G25. Fine dextran gel, Pharmacia.

Apparatus

Automatic amino acid analyser. BioCal BC 200 is a suitable apparatus (Note 1), column length 70 cm, internal diameter 0.9 cm.

Procedure
1. Depending on what type of analysis is carried out (acid-stable amino acids, sulphur amino acids, or tryptophan), load sample and standard solutions into separate loops of the sample injector, where 1 ml is retained in each separate loop (Note 3).
2. Operate the apparatus according to the conditions given in Table 1.
3. Set the programme scheme for automatic analysis according to the given schemes (Fig. 6) and run the apparatus according to the instructions.

TABLE 1. Operating conditions.

	Stable amino-acids	Sulphur amino-acids	Tryptophan
Resin/gel height in columns (cm)	56	54	55
Resin type	Bio-Rad A6	Bio-Rad A5	Sephadex G25
Temperature (°C)	56	44	30
Buffers/Bases	A, B. C; 4N NaOH	D; 0.3N LiOH	E
Buffer flow rate (ml/hr)	100	80	120
Ninhydrin flow rate (ml/hr)	50	40	60

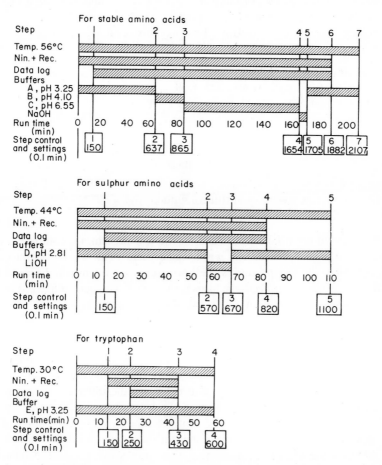

FIG. 6. Programme schemes for automatic analysis.

Calculation

Measure the areas of the recorded peaks (Note 4) either by the use of the data logger or by triangulation. Determine the amount of amino acid in the sample by comparison with the areas of sample and standard. Two or three standard runs should be performed with each batch of ninhydrin reagent.

Notes

1. The instruction given in this procedure is drawn up for the use of Bio-Cal BC 200 apparatus. If another apparatus is available follow the instructions of that particular apparatus.
2. If this buffer containing ethanol is kept in the warm over a period of time, some ethanol will evaporate, thus reducing its efficiency.
3. Normally, for each set of samples using the same batch of ninhydrin, two or three standard runs are performed.
4. Figure 5 gives an illustration of a chromatogram.

References

Spackman, D. H., Stein, W. H. and Moore, S. (1958) *Analyt. Chem.* **30**, 1190.

Moore, S. (1958) *J. Biol. Chem.* **223**, 1359.

Slump, P. and Schreuder, H. A. W. (1969) *Analyt. Biochem.* **27**(1), 182.

3.5 Estimation of Protein Quality by Chemical Score (Mitchell and Block Method)

Application

All proteins, as a semi-quantitative guide only.

Principle

The levels of essential amino acids in the test protein are compared with those in a reference protein (e.g., whole egg protein). The essential amino acid whose level in the test protein is the lowest compared with the levels in the reference sample on a percentage basis is identified. The percentage reduction for this amino acid is then used to calculate the chemical score.

Procedure

1. Determine by experiment or by reference to reliable literature the amount of essential amino acids (g per 100 g of protein) in the sample under test.
2. Compare these values with those of the essential amino acids in the reference protein (whole egg) and calculate the difference between the protein under test and the reference sample.
3. Express the difference for each essential amino-acid as a percentage of the amount of that essential amino acid in the reference protein sample.
4. Identify the essential amino acid that shows the greatest percentage reduction compared with the reference.

TABLE 2. Typical data and calculations

Essential amino acids (EAA)	EAA in reference protein (whole egg) (%)	EAA in casein (%)	EAA in wheat protein (%)	Difference as a % of EAA in reference for casein	Difference as a % of EAA in reference for wheat protein
	a	b	c	$\dfrac{b-a}{a} \times 100$	$\dfrac{c-a}{a} \times 100$
Leucine	8.9	9.8	6.9	+10	−22.5
Isoleucine	5.8	6.6	4.3	+18.8	−25.8
Cysteine and methionine	5.7	3.7	4.6	−35.1	−19.3
Valine	7.4	8.7	5.3	+17.6	−25.7
Tryptophan	1.5	1.5	0.9	0.0	−40.0
Phenylalanine	5.6	5.4	5.4	−3.6	−3.6
Lysine	6.7	9.4	2.8	+36.2	−58.3
Histidine	2.1	3.0	2.2	+40.9	+4.9
Threonine	5.0	4.6	2.8	−8.0	−44.0
Tyrosine	—	5.0	2.5	—	—

(1) For casein essential amino acid showing greatest deficit over reference protein is cysteine plus methionine; i.e. −35.1, thus chemical score = 100 − 35.1 = 64.9.

(2) For wheat protein essential amino acid showing the greatest deficit over the reference protein is lysine; i.e. −58.3, thus chemical score wheat protein = 100 − 58.3 = 41.7.

5. Subtract this percentage from 100 to give the chemical score.

Comparison of Chemical Score values with Biological values

	Chemical score	Biological value
Whole egg	100	100
Casein	65	68
Wheat	42	50

Note

More complicated procedures for calculating chemical score have been suggested based on the difference between 100 and the sum of all the potentially limiting essential amino acids in the test sample expressed as a percentage of the total essential amino acids in the test sample (FAO/WHO methods of 1955 and 1963). Despite the added complication of these calculations, they do not appear to give values that correlate significantly better with biological value.

References

Mitchell, H. H. and Block, R. J. (1946) *J. Biol. Chem.* **163**, 599.
Rippon, W. P. (1959) *Brit. J. Nutr.* **13**, 243.
F.A.O. Protein Advisory Group (1973). Bulletin III (2), p. 6.

Carbohydrates

4.1 Total Available Carbohydrate (Manual Clegg Anthrone Method)

Application
The method is applicable to all types of foodstuffs.

Principle
The material is digested with perchloric acid. Hydrolysed starches together with soluble sugars are determined colorimetrically by the anthrone method and expressed as glucose.

Reagents
Perchloric acid, 52%. Add 270 ml of perchloric acid (sp. gr. 1.70) to 100 ml of water. Cool before use.

Sulphuric acid. Carefully add 760 ml of sulphuric acid (sp. gr. 1.84) to 330 ml of water. Cool before use.

Anthrone reagent. Make up sufficient 0.1% anthrone in the above sulphuric acid for the day's requirements. Make fresh every day.

Glucose standard solution. Dissolve 100 mg glucose in 100 ml distilled water.

Glucose dilute standard solution. Dilute 10 ml of strong standard to 100 ml with water (1 ml $\equiv$ 0.1 mg glucose).

Apparatus
Filter papers. Whatman No. 542 or Schleicher and Schüll No. 150.

Spectrophotometer. For manual determination. Suitable for measurement at 630 nm.

Procedure

(*a*) *Extraction*:
  1. Weigh, to the nearest mg, 1.0 g of dry sample or 2.5 g of wet sample containing 60–300 mg total available carbohydrate.

2. Transfer, quantitatively, to a graduated 100 ml stoppered measuring cylinder.
3. Add 10 ml of water and stir with a long glass rod to disperse the sample thoroughly.
4. Add 13 ml 52% perchloric acid reagent by measuring cylinder.
5. Stir frequently with the glass rod for at least 20 minutes.
6. Wash the rod down with water and dilute the contents to 100 ml.
7. Mix and filter into a 250 ml graduated flask.
8. Wash the measuring flask with water and transfer the washings to the graduated flask.
9. Dilute to the mark with water and mix thoroughly.

(b) *Determination*:
1. Dilute 10 ml of the sample extract (a.9) to 100 ml with water.
2. Pipette 1 ml of diluted filtrate into a test tube.
3. Pipette out duplicate blanks using 1 ml of water.
4. Pipette out duplicate standards using 1 ml of dilute glucose.
5. Pipette rapidly to all the tubes 5 ml of freshly prepared anthrone reagent.
6. Stopper all tubes and mix the contents thoroughly.
7. Place in a boiling water bath for exactly 12 minutes.
8. Cool quickly to room temperature.
9. Transfer the solutions to 1 cm glass cuvettes.
10. Read the absorbance of the samples and standards at 630 nm against the reagent blanks (Note 1).

Calculation

Let: Weight (g) of sample $= W$
Absorbance of dilute standard $= a$ (Note 2)
Absorbance of dilute sample $= b$
Then: Total available carbohydrate (as % glucose) $= (25 \times b)/(a \times W)$

Notes
1. The green colour is stable for at least 2 hours.
2. The graph is a straight line over the range 0–0.15 mg of glucose (manual) or 0–1.5 mg of glucose (automated).

Reference
Clegg, K. M. (1956). *J. Sci. Food Agric.* **7**, 40.

4.2 Total Available Carbohydrate (Automated Clegg Anthrone Method)

Application
The method can be applied to all types of food products that contain potato

starch, glucose, fructose, and sucrose. The methods can also be applied if only lactose is present. Other reducing sugars present besides those mentioned above interfere, and other reducing components may also interfere.

Principle

The material is digested with perchloric acid, and solubilised starches together with soluble sugars are colorimetrically determined by the anthrone method using a continuous flow analyser. The molar extinction coefficient of the different sugars depends on the reaction time and temperature. Under the described conditions potato starch, glucose, fructose, and sucrose have the same molar extinction coefficient at 104°C. The molar extinction coefficient of lactose is about 150% of that of glucose or fructose.

Reagents and standards

Perchloric acid solution, 52%. Dilute 540 ml of perchloric acid (sp. gr. 1.70) with 200 ml of water.

Anthrone reagent. Solution I: dissolve 1.0 g of anthrone and 1 g of thiourea (NH_2CSNH_2) in 100 ml sulphuric acid (sp. gr. 1.84). Solution II: dilute 660 ml sulphuric acid (sp. gr. 1.84) with 330 ml of water and cool. Mix solutions I and II (Note 1).

Standard glucose solution. Dissolve 50, 100, and 150 mg of glucose in 100 ml of distilled water.

Standard fructose solution. Dissolve 50, 100, and 150 mg of fructose in 100 ml of distilled water (Note 2).

Apparatus

Electric stirrer. Heidolph type E60 or other suitable stirrer with stainless steel rod.

Electric oven or water bath, 50°C.

Continuous flow analyser for automatic analyses. Equipped with appropriate accessories (see Fig. 7).

Procedure

(*a*) *Extraction*:
1. Weigh 0.5 g of dry sample or 2 g of wet sample in a 100 ml polypropylene stoppered flask or tube.
2. Add 15 ml of distilled water.
3. Add 20 ml of perchloric acid solution (52%).
4. Place the flask at 50°C in an oven or water bath for 15 minutes.
5. Remove the flask from the oven and place the rod of an electric stirrer in the solution.

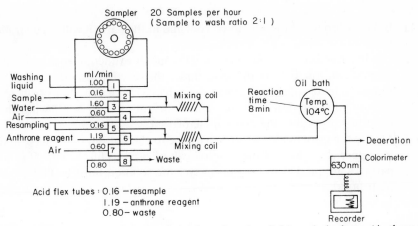

FIG. 7. Set-up for automatic determination of total available carbohydrates (Anthrone method).

6. Mix the warm solution thoroughly for 5 minutes.
7. Transfer the solution quantitatively to a 250 ml volumetric flask.
8. Make up to the mark with distilled water and mix (Note 3).
9. Stand to allow solid particles to settle.

(b) Automatic determination:
1. Set up the system, according to Fig. 7.
2. Connect the reagent tubes with the reagent flask.
3. Start the automatic system and wait until the stream passes the flow cell.
4. Fill the cups on the sample plate with standard and sample solutions (Note 4).
5. Start the sampler and run until all the samples have passed the flow cell.

Calculation

Read the optical value of samples and standard solutions. Prepare a calibration curve with, on the x-axis, the glucose concentration expressed as mg glucose per 100 ml solution (50, 100, and 150 mg resp.), and on the y-axis the optical density values. Read the glucose values of the sample from the calibration curve.

Let: This value (mg glucose/100 ml) $= a$
Weight (g) of sample $= W$
Dilution factor $= F$ (Note 3)

Then: Total available carbohydrate content (expressed as % glucose)
$$= \frac{a \times F}{4 \times W}$$

Total available carbohydrate content (expressed as % starch)
$$= \frac{a \times F \times 0.9}{4 \times W}$$

134 The Analysis of Nutrients in Foods

Notes
1. If the anthrone solution is stored cool and in the dark, it will keep for 48 hours.
2. Use this fructose solution to check if the molar extinction coefficients of glucose and fructose are the same under the reaction conditions.
3. If the 250 ml of solution contains more than 375 mg of carbohydrate (as glucose) the solution must be diluted before analysis. Following the described procedure the dilution factor = 1.
4. Mostly the insoluble particles precipitate to the bottom of the flask and the upper clear layer of the solution can be used to fill the cup. If solid particles are present in the upper layer the solution must be filtered or centrifuged.

4.3 Starch and Total Low Molecular Weight Sugars (Luff–Schoorl Method)

Application
The method is applicable to food products and ingredients containing low molecular weight sugars and natural or modified starches. Several thickening agents and glycogen interfere with the method (Note 1).

Principle
The sugars are extracted from the food product by hot ethanol, hydrolysed in aqueous solution with hydrochloric acid and determined titrimetrically as glucose. Starches are isolated from the residue of the hot ethanolic extraction after treatment with potassium hydroxide solution. On hydrolysis the starch is converted to glucose which is then determined titrimetrically.

Reagents
Sodium hydroxide solution, 300 g/l.
Hydrochloric acid solution, 1.0N.
Ethanolic potassium hydroxide solution. Dissolve 50 g of potassium hydroxide in 800 ml of 95% (v/v) ethanol and dilute with ethanol to 1000 ml.
Deproteinising solution I. Dissolve 106 g of potassium ferrocyanide $[K_4Fe(CN)_6.3H_2O]$ in water and dilute to 1000 ml.
Deproteinising solution II. Dissolve 219 g of zinc acetate $[Zn(CH_3COO)_2.2H_2O]$ in water. Add 30 ml of glacial acetic acid and dilute with water to 1000 ml.
Copper reagent. Weigh 75.0 g of anhydrous sodium carbonate (Na_2CO_3), 48.9 g of sodium hydrogen carbonate $(NaHCO_3)$, 70.0 g of trisodium citrate dihydrate $(Na_3C_6H_5O_7. 2H_2O)$, and 25.0 g of powdered copper sulphate pentahydrate $(CuSO_4.5H_2O)$. Mix the salts in a beaker and dissolve in about

800 ml of cold water, stirring continuously. Make up to 1000 ml with water. If after standing for one day turbidity or precipitate has formed, decant or filter the solution. The pH after $1 + 49$ dilution with freshly boiled water should be 10.0 ± 0.1.

Bromothymol blue solution, 10 g/l in ethanol 95 % (v/v).

Starch indicator solution. Add a mixture of 10 g of soluble starch, 10 mg of mercury(II) iodide (as a preservative), and 30 ml of water to 1 litre of boiling water. Continue boiling for 3 minutes and cool.

Sodium thiosulphate standard solution, 0.1 N. Dissolve in 1000 ml of freshly boiled and subsequently cooled water, 25 g of sodium thiosulphate ($Na_2S_2O_3$) and add 0.2 g of sodium carbonate ($Na_2CO_3.10H_2O$). Allow the solution to stand for one day before standardising.

Ethanol, 80 % (v/v).

Potassium iodide solution, 100 g/l. Dissolve 10 g of potassium iodide in water, and dilute to 100 ml. Store the solution in a dark bottle.

Hydrochloric acid solution, 6N. Dilute 100 ml of concentrated hydrochloric acid (sp. gr. 1.18) with 85 ml of water.

Potassium thiocyanate solution, 200 g/l. Dissolve 20 g of potassium thiocyanate (KCNS) in water and dilute to 100 ml.

Apparatus

Water bath. $70 \pm 2\,°C$ and boiling.

Asbestos plate with a circular hole, fitting a conical flask of 250–300 ml.

Condenser, air-cooled.

pH meter.

Centrifuge.

Procedure

(*a*) *Extraction of sugars*:
1. Weigh not more than 25 g of sample into a 250 ml centrifuge bottle.
2. Add 150 ml of 80% ethanol.
3. Cover bottle with watch glass.
4. Heat on a steam bath, with occasional stirring for 1 hour.
5. Centrifuge for 10 minutes at 3000 r.p.m.
6. Decant the supernatant into an evaporating basin.
7. Wash the residue with 50 ml of 80% ethanol.
8. Centrifuge for 10 minutes at 3000 r.p.m.
9. Decant and combine the supernatant solutions (keep the residue for starch determination, see 'Isolation of starch').
10. Evaporate to a small volume until all the alcohol has been removed.
11. Transfer the residue with water to a 100 ml graduated flask.

12. Add 3 ml of deproteinising solution I and, after mixing, 3 ml of deproteinising solution II, then dilute to 100 ml.
13. Mix and filter through a filter paper.

(b) *Acid hydrolysis of sugars*
  1. Pipette an aliquot (1 ml) of solution from instruction a.13 containing not less than 40 mg and not more than 240 mg of glucose into a test tube.
  2. Add a quantity of concentrated hydrochloric acid, so that the final concentration in the solution is approximately 1N.
  3. Place the tube in a water bath at 70°C ($\pm$2°C).
  4. Heat for exactly 10 minutes.
  5. Cool and transfer the solution to a 100 ml graduated flask.
  6. Neutralise with 40% sodium hydroxide using bromothymol blue indicator and taking care not to exceed pH 6.5.
  7. Dilute to volume with water. Immediately prior to use add 1–2 drops of 40% sodium hydroxide to make the solution alkaline to bromothymol blue.

(c) *Determination of sugars:*
  1. Pipette a 25 ml aliquot from instruction (b)7 containing 10–60 mg of glucose into a 250 ml conical flask.
  2. Pipette 25 ml of copper reagent into the flask and add several anti-bumping granules.
  3. Connect the flask to a condenser.
  4. Stand the flask on a wire gauge.
  5. Bring the contents of the flask to the boil in about 2 minutes and then continue to boil for exactly 10 minutes.
  6. Cool the flask quickly to room temperature.
  7. Add 5 ml of potassium iodide solution, followed carefully but quickly with 20 ml of 6N hydrochloric acid.
  8. Add 10 ml of thiocyanate solution.
  9. Titrate the liberated iodine with standard sodium thiosulphate. When the solution becomes pale yellow add 1 ml of starch indicator and continue the titration until the blue colour disappears.
  10. For a blank determination carry out instructions (c)2–9 using 25 ml of water instead of 25 ml of glucose containing filtrate.

(d) *Isolation of starch*:
  1. Transfer the residue [see instruction (a)9] to a 500 or 600 ml beaker. (If the quantity of starch in this portion of sample is more than 500 mg, reduce the amount of the sample accordingly.) The minimum amount of starch is 80 mg(Note 3).

2. Add, while stirring by means of a glass rod, 300 ml of hot ethanolic potassium hydroxide solution and cover the beaker with a watch glass.
3. Heat on the boiling water bath for 1 hour with occasional stirring.
4. Decant the solution through the filter paper and then wash the starch quantitatively on to the filter paper using hot ethanol 80 % (v/v) with the aid of rubber-tipped glass rod. Keep the filter moist (Note 4).
5. Immediately loosen the precipitate from the paper by means of a glass rod.
6. Pierce a hole in the filter paper and wash the starch through it into a 250 ml beaker, using 100 ml of hot 1N hydrochloric acid solution.
7. Cover the beaker with a watch glass and stand it in the boiling water bath for $2\frac{1}{2}$ hours, stirring occasionally with a glass rod.
8. Cool the solution and neutralise by adding the sodium hydroxide solution dropwise, taking care not to exceed pH 6.5; check this with a pH-meter.
9. Transfer the mixture to a 200 ml graduated flask with the aid of water, add 3 ml of deproteinising solution I and, after mixing, 3 ml of deproteinising solution II and dilute to 200 ml.
10. Mix and filter through a fluted filter paper. Immediately before pipetting an aliquot portion for the next stage, make the filtrate alkaline to bromothymol blue by adding 1–2 drops of the sodium hydroxide.
11. Proceed as given in instructions (c)1–10 above.

Calculation
(a) Total low molecular weight sugars as glucose:
     Let: Weight (g) of sample $= W$
             Volume (ml) of standard thiosulphate solution
             needed for the blank $= V_0$
             Volume (ml) of standard thiosulphate solution
             needed for the determination $= V$
             Normality of the standard thiosulphate solution $= T$
     Calculate the amount of (ml) of 0.1000N thiosulphate solution as follows:
$$(V_0 - V) \times T \times 10$$
Find in the conversion table (Table 3) the corresponding amount of glucose.
     Let: This amount (mg) $= g$
     Then: Sugar content (%) $= [g/(W \times a)] \times 40$
             expressed as glucose
(*b*) Starch content:
Calculate the amount (ml) of 0.1000N thiosulphate solution as follows:
$$(V_0 - V) \times T \times 10.$$
Find in the conversion table (Table 3) the corresponding amount of glucose.
     Let: This amount (mg) $= g$
Then: The starch content (%) $= (g \times 800 \times 0.9)/W$ (0.9 is the factor used to convert glucose equivalents to starch).

Notes

1. Several thickening agents and glycogen interfere with the method, viz.,

	Recovered as starch (%)
Glycogen	100
Locust bean flour	100
Carboxymethylcellulose	∼27
Alginate	∼15
Carrageenan	∼25
Agar	∼47

2. The total volume of liquid should always be 50.0 ml.
3. If the approximate starch content is unknown, a trial analysis should be made first.
4. In some cases centrifugation may be more advantageous than filtration.

TABLE 3. For converting the differences in millilitres of 0.1N thiosulphate needed for the filtrate and the blank determination $(V_0 - V) \times 10 \times T$ to glucose.

ml of 0.1N thio $(V_0-V) \times 10 \times T$	Glucose (mg)	Δ	ml of 0.1N thio $(V_0-V) \times 10 \times T$	Glucose (mg)	Δ
1	2.4		13	33.0	
		2.4			2.7
2	4.8		14	35.7	
		2.4			2.8
3	7.2		15	38.5	
		2.5			2.8
4	9.7		16	41.3	
		2.5			2.9
5	12.2		17	44.2	
		2.5			2.9
6	14.7		18	47.1	
		2.5			2.9
7	17.2		19	50.0	
		2.6			3.0
8	19.8		20	53.0	
		2.6			3.0
9	22.4		21	56.0	
		2.6			3.1
10	25.0		22	59.1	
		2.6			3.1
11	27.6		23	62.2	
		2.7			
12	30.3				
		2.7			

4.4 Starch (Enzymic Method)

Application
This method gives good results for products containing natural starch. Products containing modified starches, especially epichlorohydrin cross-linked starches, give low recoveries.

Principle
Low molecular weight sugars are first removed with hot 80% ethanol. If much fat and proteinaceous material is present, the sample is then treated with hot ethanolic potassium hydroxide and washed with 80% ethanol. The starch in the residue is gelled and then incubated with amyloglycosidase enzyme in a buffer at pH 4.5. The enzyme degrades the starch to glucose which is measured by the glucose oxidase method.

Reagents
Sulphuric acid, 8N. Add 112 ml of concentrated sulphuric acid (sp. gr. 1.84) to 350 ml of water, carefully and with cooling. When cool dilute to 500 ml.
Acetic acid, 0.2N. Dilute 12 ml of acetic acid to 1 litre.
Sodium acetate, 0.2N. Weigh 27.2 g of sodium acetate trihydrate ($CH_3COONa.3H_2O$) and make up to 1 litre.
Enzyme reagent. Weigh 0.5 g of amyloglucosidase (Grade II from *Rhizopus* genus mould, supplied by Sigma Chemical Co.) and dissolve in 50 ml of buffer solution consisting of a mixture of 20 ml of 0.2N sodium acetate and 30 ml of 0.2N acetic acid (pH 4.5). Allow insoluble material to settle. Prepare the reagent just before use.
Glucose oxidase peroxidase-chromogen reagent. 0.1N Acetate buffer pH 5.5. Take 45 ml of 0.2N sodium acetate, add 5 ml of 0.2N acetic acid, mix, and dilute to 100 ml. Weigh 20 mg of purified, crystalline *o*-dianisidine di-hydrochloride (Sigma Chemical Co.) into a tube and dissolve in 2 ml of water and 1 ml of ethanol. Into another tube weigh 2 mg of horseradish peroxidase (Type II, Sigma Chemical Co.) and dissolve in 5 ml of buffer pH 5.5; then add 0.3 ml of glucose oxidase (from *Aspergillus niger*, Type V, 1000 units/ml, Sigma Chemical Co.). Transfer the solutions in the two tubes to a 100 ml brown glass reagent bottle, washing and mixing with the pH 5.5 buffer. Finally add all the buffer (100 ml), stopper, and mix. Prepare the reagent just before use.
Ethanol, 80% v/v.
Glucose standards. Weigh 100 mg of glucose, dissolve in water, and make up to 1 litre. Pipette 10, 20, 40, and 60 ml aliquots in 100 ml graduated flask and make up to the mark. These standard solutions contain 10, 20, 40, and 60 μg of glucose per ml respectively.

Apparatus
    Centrifuge.
    Steambath.
    Spectrophotometer or colorimeter. Suitable to measurement at 526 nm.
    Glass cuvette. Optical pathway 1 cm.

Procedure
 1. Weigh sufficient sample to contain 0.1–0.2 g of starch into a 250 ml centrifuge bottle.
 2. Extract the sugars with 80% ethanol as described in methods a.1 to a.9.
 3. If the sample is non-fatty and low in protein (i.e. cereals) proceed to instruction 11.
 4. Add 150 ml of hot ethanolic potassium hydroxide solution to the residue from method 4.3 instruction a.9.
 5. Cover the bottle with a watch glass.
 6. Heat on a steam bath with occasional stirring for 1 hour.
 7. Centrifuge for 10 minutes at 3000 r.p.m.
 8. Decant and discard the supernatant liquid.
 9. Wash the residue with 50 ml hot 80% ethanol.
 10. Centrifuge for 10 minutes at 3000 r.p.m. and discard the supernatant.
 11. To the residue in the centrifuge bottle add 20–30 ml water and heat the uncovered bottle in the oven at 100°C for 1 hour (to remove traces of ethanol.
 12. Cover the mouth of the bottle with aluminium foil and continue heating at 100°C for a further 2 hours to gel the starch.
 13. Remove bottle, allow to cool to below 40°C, add 20 ml of 0.2N sodium acetate followed by 30 ml 0.2N acetic acid, and mix.
 14. Add 2.0 ml of freshly prepared amyloglucosidase enzyme reagent and mix.
 15. Add 2 drops (0.05 ml) of toluene, cover the bottle with the foil, and incubate overnight (16 hours) at 55–58°C.
 16. Allow to cool and transfer contents of the bottle to a 100 ml graduated flask, washing out the bottle with a little water.
 17. Dilute the contents to 100 ml, mix well, and allow the contents to settle.
 18. Pipette 2.0 ml of the clear supernatant into a 100 ml graduated flask, dilute to 100 ml with water, and mix.
 19. Into glass stoppered tubes pipette (in duplicate) 1.0 ml of this solution.
 20. Add 2.0 ml of the glucose oxidase–peroxidase–chromogen reagent. Stopper and mix.
 21. Incubate the tubes at 37°C for 1 hour.
 22. Add 5.0 ml of 8N sulphuric acid to each tube, stopper, mix, and cool to room temperature.

23. Pipette 1 ml of glucose standards (10, 20, 40, and 60 μg glucose/ml) and a
    blank using 1 ml of water into the tubes and run concurrently with the
    samples. Proceed according to instructions 20–22.
24. Transfer the solutions to 1 cm glass cuvettes and measure the absorbance
    at 526 nm.

Calculation

(*a*) *Calibration curve:*
Correct the absorbance of the standards for the blank value. Prepare a
calibration curve with concentration of glucose in μg/ml on the x-axis and the
corrected absorbances on the y-axis.

(*b*) *Starch content:*
Correct the absorbance value of the sample for the blank value. Find from
the calibration curve, using the corrected absorbance value, the glucose
concentration.

Let: This value (μg/ml) $= C$
       Weight (g) of sample $= W$
Then: Starch content (%) $= 0.45 \times (C/W)$

Reference
McRae, J. C. (1971). *Planta* (Berlin) **96**, 101.

4.5 Reducing Sugars and Sucrose (Automated Picric Acid Method)

Application
   The method can be applied to all types of food products that contain glucose
and/or fructose and sucrose in concentrations less than 10%. The method can
also be applied if only lactose is present. Other reducing sugars present besides
those mentioned are also measured.

Principle
   Sugars are extracted from the food product with ethanol. After evaporation
of the ethanol the residue is dissolved in water. Using a continuous flow system
the extract is mixed with invertase which converts sucrose into glucose and
fructose. After dialysis the sugars in solution are reacted with picrate to give
picramide which is measured colorimetrically. The procedure can be adapted to
measure total sugars, reducing sugars and lactose in products containing only
that sugar (Note 1).

A. EXTRACTION OF THE SUGARS

Reagents

Alcohol, 96% w/v and 80% v/v.

Apparatus

Centrifugal filter.
Filter paper strips. Schleicher and Schüll No. 595 or Whatman No. 1.
Rotary vacuum evaporator.
Waterbath. Temperature about 60°C.
Steam bath.

Procedure

1. Weigh 10 g of the homogenised product into a 100 ml beaker.
2. Add 50 ml of 96% alcohol and mix thoroughly.
3. Place the covered beaker on a steam bath and heat sample to just below boiling. Heat for 1 hour, stirring occasionally.
4. Filter the slurry using the centrifugal filter (Note 2). Collect the filtrate in a 100 ml graduated flask.
5. Wash the beaker three times with 5 ml of 80% alcohol and transfer the washings into the centrifuge.
6. Wash the slurry in the centrifuge with 80% alcohol until the volume collected is nearly 100 ml.
7. Cool to room temperature, make up to the mark, and mix.
8. Pipetter 20 ml (Note 3) of the alcoholic solution into a 100 ml pear-shaped vessel.
9. Evaporate the alcoholic solution until dry using the rotary vacuum evaporator.
10. Pipette 5 ml of water to the dried residue.
11. Swirl the flask and contents in a waterbath at 60°C for about 1 minute.
12. Cool to room temperature and filter through a cotton-wool filter.

B. DETERMINATION OF SUGARS

Reagents

Picric acid solution, 1.2%.
Picric acid solution, 1%. Dilute 333 ml of 1.2% picric acid solution with water to 400 ml.
Sodium hydroxide solution, 14%. Dissolve 7 g of sodium hydroxide in water and make up to 50 ml.
Sodium hydroxide picrate solution. Add 50 ml of 14% sodium hydroxide to 400 ml of 1% picric acid solution. Dilute with water to 500 ml. Prepare this solution freshly every day.

Buffer solution, pH 4.5. Dissolve 11.7 g of sodium acetate trihydrate in 500 ml of water. Add 6.5 ml of glacial acetic acid and dilute to 1 litre.

Invertase. Store the invertase at 5°C.

Invertase solution. Dissolve 20 mg of invertase in 100 ml of buffer solution. Prepare this solution fresh every day.

Sucrose.

Sucrose standard solutions. Prepare solutions containing 2, 4, 8, 20, and 80 mg/ml by dissolving 0.200, 0.400, 0.800, 2.0, and 8.0 g respectively in water and diluting each to 100 ml.

Apparatus

Continuous flow analyser for automatic analysis with appropriate acccessories (see Fig. 8).

Procedure

1. Assemble the apparatus as in Fig. 8. Pump the reagents through until the system is equilibrated.
2. Fill the sampler cups with standard sucrose solutions followed by sample solutions. The range of standards should be repeated after every 12–15 samples.

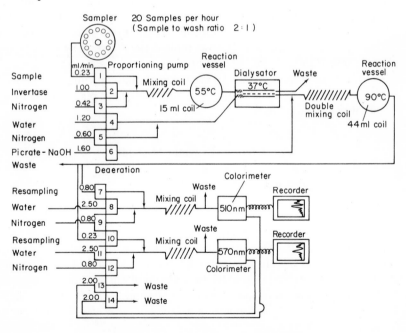

FIG 8. Set-up for automatic sugar determination.

3. Run the system until all samples have passed the flow cells of the colorimeters.

Calculation
1. Read the absorbance of blanks, samples, and calibration solutions from the recorder paper.
2. Correct the absorbances of standards and samples for the blank value.
3. Prepare if necessary two calibation curves, one for the low range (510 nm), one for the high range (570 nm).
4. Read the sugar values (expressed as g of sucrose per 100 ml) of the sample from the calibration curves.

Let: This value (g) $= a$
        Weight (g) of product $= W$
Then: The sugar content (%) expressed as sucrose $= (a \times 25)/W$
      or as invert sugar $= (a \times 25 \times 1.05)/W$

If a 40 ml aliquot is used at instruction a.8, then the sugar content (%) expressed as sucrose $= (a \times 12.5)/W$
or as invert sugar $= (a \times 12.5 \times 1.05)/W.$

Notes
1. The method described is for total sugars. If reducing sugars only are required, replace the invertase by water. The difference between the reducing sugar result and the total sugar result gives the sucrose content. In samples containing only lactose, replace the invertase with water and use lactose standards for calibration.
2. If a centrifugal filter is not available a conventional centrifuge can be used.
3. If the sugar content of the sample is smaller than 0.5 % pipette 40 ml into a 200 ml pear-shaped flask.

4.6 Lactose (Enzymic Method)

Application
The method is generally applicable to foodstuffs.

Principle
Lactose is converted into glucose and galactose by β-galactosidase (β-Gal). The galactose is then oxidised in the presence of the enzyme galactose dehydrogenase (Gal-DH) to galactono-lactone and NADH by nicotinamide adenine dinucleotide (NAD). NADH, the amount of which is equivalent to the lactose present in the sample, is measured at 340 nm or 366 nm.

$$\text{Lactose} + H_2O \xrightarrow{\beta\text{-Gal}} \text{glucose} + \text{galactose}$$

$$\text{Galactose} + NAD^+ \xrightleftharpoons{\text{Gal-DH}} \text{galactono-lactone} + NADH + H^+$$

Reagents

Deproteinizing solution 1. Dissolve 106 g of potassium ferrocyanide trihydrate $[K_4Fe(CN)_6.3H_2O]$ in water and dilute to 1 litre.

Deproteinizing solution 2. Dissolve 219 g of zinc acetate dihydrate $[(CH_3COO)_2Zn.2H_2O]$ in water, add 30 ml of glacial acetic acid, and dilute to 1 litre.

Sodium hydroxide solution, 1N. Dissolve 40 g of sodium hydroxide in water and dilute to 1 litre.

Buffer solution. Dissolve 4.80 g of disodium hydrogen phosphate, 0.86 g of sodium dihydrogen phosphate monohydrate, and 0.20 g of magnesium sulphate heptahydrate in 200 ml of water. Check that the pH is 7.5. Stable indefinitely at approximately $+4°C$.

NAD solution. Dissolve 50 mg of nicotinamide adenine dinucleotide (NAD) in 5 ml of water. Stable for 4 weeks at approximately $+4°C$.

Gal-DH suspension. β-Galactose dehydrogenase (Gal-DH), 5 mg/ml. Use undiluted. Stable for 1 year at approximately $+4°C$.

β-Galactosidase suspension. β-galactosidase (β-Gal), 5 mg/ml. Use undiluted. Stable for 1 year at approximately $+4°C$.

Lactose standard solution. Dissolve 100 mg of lactose in water and dilute to 1 litre. Prepare just before use.

Apparatus

Spectrophotometer. Suitable for measurement at 340 or 366 nm.
Glass cuvettes. Optical pathway 1 cm.
High speeds macerator. The Ultra-Turrax is suitable.
Glass fibre filter paper. Whatman GF/A.

Procedure

(a) *Sample extraction*:

1. Weigh, to the nearest mg, 5 g of sample into a 100 ml beaker.
2. Add 25 ml of warm water and macerate to suspend the sample in the water.
3. Pour the suspension into a 200 ml graduated flask.
4. Add 2.5 ml of deproteinising solution 1 and mix.
5. Add 2.5 ml of deproteinising solution 2 and mix.
6. Add 10 ml of 1N sodium hydroxide, mix, and dilute to volume.
7. Filter through glass fibre filter paper.

(b) *Determination of lactose*

N.B. The extract prepared in instruction a.7 should not contain more than 0.400 g/litre lactose, if measured at 366 nm, or 0.200 g/litre if measured at 340 nm.

1. Pipette into 1 cm glass cuvettes:

	Blank	Standard	Sample
Buffer solution	3.00 ml	3.00 ml	3.00 ml
NAD solution	0.10	0.10	0.10
Redistilled water	0.20	—	—
Lactose standard	—	0.20	—
Sample	—	—	0.20

2. Pipette into each cuvette 0.02 ml of Gal-DH suspension.
3. Mix, wait approximately 30 minutes, and read the absorbances of the standard (E_{1st}) and sample (E_{1s}) against the blank.
4. Pipette into each cuvette 0.02 ml of β-galactosidase suspension.
5. Mix, wait approximately 30 minutes, and read the absorbances of the standard (E_{2st}) and sample (E_{2s}) gainst the blank.

Calculation

Let: Weight of sample $= W$

Then: lactose content (%) $= \dfrac{2.0}{W} \times (E_{2s} - E_{1s})/(E_{2st} - E_{1st})$

Notes

1. Reagents supplied by Boehringer are suitable. When ordering reagents, specify the enzymic method to be used.

4.7 Sugar Composition (High Performance Liquid Chromatographic Method)

Application

The method can be applied to the mono- and di-saccharide constituents of food products.

Principle

An extract of the product or a hydrolysate is applied to an ion exchange column equilibrated with ethanol–water. The mono- and/or di-saccharides are separated by elution with ethanol–water. This chromatographic separation can be most rapidly and quantitatively done by applying high performance liquid chromatography. For the detection of the sugars in the eluate a moving wire detector (Pye LCM2) is recommended. Three column systems are required to most effectively carry out routine mono- and di-saccharide analysis. Lithium columns are preferred for disaccharides, trimethylammonium columns for monosaccharides (especially hydrolysates), and sulphate columns for glucose, fructose, and sucrose mixtures.

Reagents (N.B. Use deionised water throughout)

Aminex A-6, ex Bio Rad, Bromley, Kent. Strongly acidic cation exchange resin converted into the lithium form.

Aminex A-6. Strongly acidic cation exchange resin converted into the trimethylammonium form.

Technicon, ex Basingstoke, Hants, type S. Strongly basic anion exchange resin (×8 cross-linked, 16–18 μ particles) in the sulphate form.

Ethanol, 96% w/v.

Ethanol–water, 85% ethanol w/w and 75% ethanol w/w.

Mono- and di-saccharides for use as standards. Rhamnose, fucose, ribose, xylose, arabinose, fructose, mannose, glucose, galactose, sucrose, maltose, lactose. Make up a mixed standard of at least three of the appropriate sugars in 75% ethanol (30 mg/ml of each sugar).

Glass beads, 100 μm diameter.

Apparatus

Liquid Chromatograph. See Fig. 9, equipped with appropriate accessories.

Injection device. Syringe or sample loop.

Detector. Pye LCM 2 detector is recommended.

Columns. Glass or precision stainless steel, 50 or 100 cm in length and 0.4 cm internal diameter; these are contained in a simple thermostatted jacket controlled from a constant-temperature bath with circulating pump. Details of columns and operating conditions are given in Table 4.

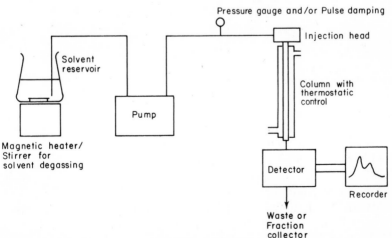

FIG. 9. Schematic diagram of liquid chromatograph.

TABLE 4. Details of Columns and Operating Conditions

Resin Type	Aminex A-6	Aminex A-6	Technicon Type S [1]
Form	Lithium	Trimethylammonium	Sulphate
Dimensions	50 × 0.4 cm i.d.	100 × 0.4 cm i.d.	50 × 0.4 cm i.d.
Solvent (w/w)	85% ethanol in water	85% ethanol in water	85% ethanol in water
Temperature	75°C	75°C	65°C
Fig.	10	11	12
	Relative retention [2]	Relative retention [2]	Relative retention [2]
Rhamnose	1.00	1.00	1.00
Fucose		1.28	1.32
Xylose	1.88	1.85	2.52 [6]
Ribose	2.48 [3]	1.45	1.45
Arabinose	2.54 [3]	2.03	2.14
Mannose	3.25 [4]	2.33	2.34
Fructose	3.25 [4]	1.96	2.50 [6]
Glucose	3.23 [4]	2.61 [5]	4.62
Galactose	4.06	2.75	4.01
Maltose	5.76	2.97	10.79
Sucrose	5.92	2.56 [5]	10.30
Lactose	8.67	4.15	11.04
Approximate time of run			
Monosaccharides	—	$1\frac{1}{2}$–2 hours	1–$1\frac{1}{4}$ hours
Disaccharides	$1\frac{1}{2}$–2 hours	4 hours	4 hours

[1] The relative retentions hold for 85% ethanol. This column gives more rapid elution of fructose, glucose, and sucrose mixtures if 75% ethanol is used, this solvent being recommended.
[2] Relative retention figures given below will vary slightly from column to column and should not be taken as absolute.
[3] Ribose and arabinose are not fully resolved.
[4] Glucose, fructose, and mannose are unresolved.
[5] Glucose and sucrose are not fully resolved and sucrose may decompose.
[6] Fructose and xylose are unresolved.

Procedure

(a) *Column packing*:
  1. Slurry the ion exchange resin in a flask with the developing solvent (85 or 75% w/w ethanol in water) and allow to equilibrate for 1 hour.
  2. Add the slurry to the column and allow to settle, aliquots being added to bring the resin to the required bed depth.
  3. Fill the final 5 cm of the column with glass beads.

(b) *Sample preparation*:
  1. Warm 2 g of sample with 40 ml of 75% ethanol in water on a steam bath for 1 hour.
  2. Separate the supernatant liquid from any solid residue by filtration and evaporate to dryness.

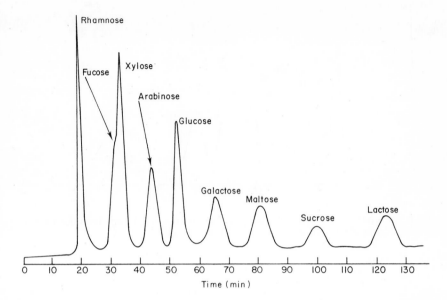

FIG. 10. Separation of mono- and di-saccharides. Column 1 m × 4 mm, Aminex A-6 lithium form; 85% w/w ethanol–water; 75°C; 1000 lb/in².

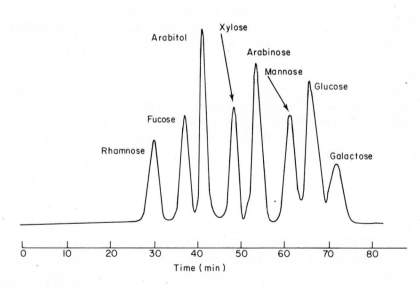

FIG. 11. Separation of monosaccharide mixture. Column 1 m × 4 mm, Aminex A-6, trimethylammonium/sodium (9:1) form; 85% w/w ethanol–water; 75°C; 500 lb/in²; 0.55 ml/min.

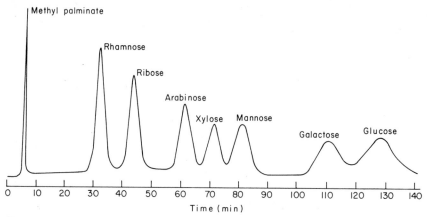

FIG. 12. Separation of monosaccharides. Column 0.5 m × 4 mm, Technicon type S, sulphate form; 85% w/w ethanol–water; 65°C; 0.5 ml/min.

3. Make up the residual solids to approximately 100 mg/ml with the column solvent (Note 1).

(c) *Chromatography*:

1. Inject by means of a syringe or by sample loop 5–50μl so that between 50 μg and 5 mg total sugar is loaded on the column. Run the apparatus according to the operating instructions (see Table 4).
2. After the sample chromatogram is complete, inject on to the column 5 μl mixed sugar standard and run as before (Note 2).

Calculation

Chromatograms of the sugar mixtures are obtained on the potentiometric recorder. Peak areas may be measured manually or using an electronic integrating device connected to the detector output. Manual peak area measurement is conveniently obtained from peak height × peak width at 0.607 of peak height.

Let: Initial sample weight (g) $= W$
      Total volume (ml) of sample extract $= V$
      Volume (μl) injected into chromatograph $= x$
      Peak area of individual sugars in sample a, b, c etc.
      Peak area of individual sugar standards $= S_a, S_b, S_c$ etc.
      Weight (μg) of individual standard injected $= w_a, w_b, w_c$ etc.

Then: Individual sugar (%) $$= \frac{a \times w_a \times V}{S_a \times x \times W \times 10}$$

Calculate similarly for b, c, etc.

Notes
1. With high fat products, it may be necessary to extract the supernatant liquid with carbon tetrachloride before the evaporation stage. Injections should be made in the column solvent but since high concentrations of disaccharides are not attainable in 85% w/w ethanol in water, it is permissible to make the solution up in a smaller volume of water.
2. When disaccharides are required, which need up to 4 hours for separation on two of the column systems, it is recommended that an internal sugar standard is added to the sample run.

References
Lawrence, J. G. and Hobbs, J. N. (1972). *J. Sci. Fd. Agric.* **23**, 45.
Lawrence, J. G. and Hobbs, J. N. (1972). *J. Chrom.* **72**, 311.

4.8 Crude Fibre

Application
This method is suitable for all foodstuffs.

Principle
A fat-free sample is treated with boiling sulphuric acid and subsequently with boiling sodium hydroxide. The residue after subtraction of the ash is regarded as fibre.

Reagents
Hydrochloric acid, 1% v/v. Add 10 ml of concentrated hydrochloric acid (sp. gr. 1.18) to water and dilute to 1 litre.
Sulphuric acid stock solution, 10% w/v. Add 275 ml of concentrated sulphuric acid (sp. gr. 1.86) to water and dilute to 5 litres.
Sulphuric acid working solution, 1.25%. Dilute 625 ml of the stock solution to 5 litres.
Sodium hydroxide stock solution, 10% w/v. Dissolve 500 g of sodium hydroxide in water and dilute to 5 litres.
Sodium hydroxide working solution, 1.25%. Dilute 625 ml of the stock solution to 5 litres.
Petroleum ether, b.p. 40–60°C.
Alcohol. Industrial methylated spirit, 95–96% v/v.
Acetone.
Antifoam. 2% silicon antifoam in carbon tetrachloride.

Apparatus
Conical flasks, 1 litre.
Buchner flasks, 1 litre.
Gooch funnels and rubber adapters.
Sintered glass crucibles, porosity No. 1.

F

Hartley or Büchner funnels.
Acid and caustic dispensers, 200 ml, automatic.
Isomantles, 1 litre.
Filter papers. Whatman No. 54 or Schleicher and Schüll No. 1573, diam
12.5 cm.
Condensers. Cold finger type.

Procedure
1. Weigh a suitable quantity of sample (usually 1 or 2 g) into a 1 litre conical
   flask (Note 1).
2. Using the dispenser, add 200 ml of 1.25 % sulphuric acid which has been
   brought to boiling point. Use the first 30–40 ml to disperse the sample.
3. Add a few drops of anti-foam and heat to boiling within 1 minute.
4. Boil gently for exactly 30 minutes under cold finger condensers. Rotate the
   flasks occasionally to mix the contents and remove particles from the side.
5. Filter the contents of the flask through a Hartley or Büchner funnel
   prepared with a wet 12.5 cm filter paper (Note 2).
6. Wash the sample back into the original flask with 200 ml of 1.25 % sodium
   hydroxide, using the automatic dispenser, measured at room temperature
   and brought to boiling point.
7. Boil for exactly 30 minutes taking the same precautions as in the earlier
   boiling treatment.
8. Transfer all insoluble matter to the sintered crucible by means of boiling
   water.
9. Wash successively with boiling water, 1 % hydrochloric acid, and boiling
   water again until acid-free.
10. Wash twice with alcohol.
11. Wash three times with acetone.
12. Dry at 100°C to constant weight.
13. Ash in a muffle furnace at 550°C for 1 hour.
14. Cool crucible in a dessicator and re-weigh.

Calculation

Let: Weight (g) of sample $= W_1$
       Weight (g) of insoluble matter $= W_2$
       Weight (g) of ash $= W_3$
Then: Fibre content (%) $= [(W_2 - W_3)/W_1] \times 100$

Notes
1. If the sample contains more than 1 % fat, extract with petroleum as
   follows:
   Add 20 ml of petroleum ether.

Stir the sample and allow to settle.
Decant the petroleum ether.
Repeat this process twice more and allow the sample to air dry.
2. If the sample does not filter within 10 minutes, repeat the determination
   using less of the material.

4.9 Acid Detergent Fibre

Application
This method is suitable for all foodstuffs.

Principle
The ground and dried sample is boiled with cetyltrimethylammonium
bromide in 1N sulphuric acid under reflux for 2 hours, and the filtered and
dried residue reported as acid detergent fibre.

Reagents
Acid detergent solution. 1% w/v cetyltrimethylammonium bromide in
H_2SO_4. Add 56 ml conc. H_2SO_4 to water and dilute to 2 litres. Dissolve 20 g of
cetyltrimethylammonium bromide in this solution.
Dekalin antifoam (Decahydronaphthalene).
Acetone.

Apparatus
Force draught oven.
Conical flasks, capacity 500 ml.
Isomantle electric heating equipment.
Cold finger condenser.
Sintered glass crucibles, porosity 1.

Procedure
1. Grind all samples in a hammer mill to pass a 1 mm sieve (16 mesh).
2. Take a sub-sample and dry overnight in a forced-draught oven at 95°C.
3. Allow samples to cool in a desiccator.
4. Weigh out, in duplicate, to the nearest mg, 1 g of the ground, dried sample
   into a 500 ml conical flask.
5. Add 100 ml cold acid detergent solution and 2 ml of dekalin antifoam.
6. Put flask into Isomantle heater and place cold finger condenser in flask.
7. Bring rapidly to the boil (3–5 min) and continue boiling gently and evenly
   under reflux for 2 hours.
8. Filter the contents of the flask, under gravity, through a previously tared
   sintered crucible.

9. Wash out the flask with hot distilled water, adding washings to the crucible.
10. Using slight suction, wash the contents of the crucible thoroughly with hot water (use approx. 300 ml of hot water in all).
11. Wash the residue with acetone and suck dry.
12. Place the crucible and contents in an oven at 95°C to dry overnight.
13. Cool in a dessicator and weigh.

Calculation

Let: Weight (g) of sample $= W_1$
      Weight (g) of residue $= W_2$
Then: Acid detergent fibre content (%) $= (W_2/W_1) \times 100$

Reference

Clancy, M. J. and Wilson, R. K. (1966). Report of the 10th International Grassland Congress, Helsinki 1966. Section 2 paper No. 22.

Lipids

5.1 Extractable Fat (Soxhlet Method)

Application

The method is generally applicable, but less precise than other methods, i.e. Weibul, Rose Gottlieb, etc.

Principle

The fat is extracted with petroleum ether from the dried residue obtained in the determination of the moisture content (Method 2.1 or 2.2). The solvent is removed by evaporation and the residue of fat is weighed.

Reagents

Petroleum ether, boiling range 40–60°C.

Apparatus

Extraction apparatus. Continuous, for example the Soxhlet type with an extraction flask of about 150 ml capacity.

Air oven.

Heating units. Water bath or steam bath.

Extraction thimbles.

Procedure

1. Dry the sample according to Method 2.1 or 2.2.
2. Transfer the dried residue to an extraction thimble (Note 1).
3. Wipe the moisture dish with small pieces of cotton wool dampened with petroleum ether and transfer the pieces to the thimble.
4. Place the thimble in the extractor and connect a weighed flask containing 100 ml petroleum ether. Connect the extractor to a reflux condenser.
5. Extract the sample, under reflux, on a water or steam bath for 2–3 hours (Note 2).

6. Evaporate the petroleum ether extract to dryness and add 2 ml acetone. Blow air gently into the flask to remove the last traces of solvent.
7. Dry the flask containing the fat residue in an air oven at 100°C for 5 minutes, cool in a desiccator, and weigh.

Calculation

Let: Weight (g) of sample before drying $= W_1$
        Weight (g) of flask without fat $= W_2$
        Weight (g) of flask with fat $= W_3$
Then: Extractable fat (%) $= [(W_3 - W_2)/W_1] \times 100$

Notes

1. Any hard lumps formed during drying should be carefully broken into small pieces before being transferred to the thimble. This will assist extraction of the fat. Fried potato products should not be dried for too long as this will tend to prevent quantitative extraction of the fat from the sample, even after 6 hours.
2. Normally 2–3 hours should ensure complete removal of the fat. To test for complete removal, disconnect the flask and replace it by another after the initital extraction time is completed. Continue the extraction for a further 1 hour, evaporate the petroleum ether, and dry the flask as before. No fat should be found in this second flask. If this is not the case, the extraction time for the particular product should be increased.

5.2 Total Fat (Weibul Method)

Application

The method is applicable to most food products except certain dairy products.

Principle

The sample is boiled in dilute hydrochloric acid to free the occluded and bound lipid fractions and is subsequently extracted with n-hexane or petroleum ether (see Note 1).

Reagents

Hydrochloric acid, 4N. Dilute 100 ml of concentrated hydrochloric acid (sp. gr. 1.18) with 200 ml of water and mix.

Extraction solvent. n-Hexane or petroleum ether, boiling range 40–60°C. For either solvent, the residue on complete evaporation should not exceed 2 mg per 100 ml.

Blue litmus paper.

Boiling chips.

Apparatus

Clock glass or Petri dish: diameter not less than 80 mm.

Extraction thimble: fat-free: Whatman Extraction Thimbles or Schleicher and Schüll No. 603.

Cotton wool: defatted.

Extraction apparatus: continuous, for example the Soxhlet type with an extraction flask of about 150 ml capacity.

Water bath.

Air oven: temperature $103 \pm 2°C$.

Fluted filter paper: qualitative, of medium speed.

Procedure

1. Weigh, to the nearest mg, 3–5 g of the sample, depending on the fat content, into a 250 ml conical flask.
2. Dry the flask of the extraction apparatus, containing some boiling chips, for 1 hour at $103 \pm 2°C$ in the drying oven.
3. Allow the flask to cool to room temperature in the desiccator and weigh to the nearest mg.
4. Add to the weighed sample 50 ml of the hydrochloric acid and cover the conical flask with a small watch glass.
5. Heat the conical flask on an asbestos wire gauze by means of a gas burner until the contents start boiling.
6. Continue boiling over a small flame for 1 hour and shake occasionally. Add 150 ml of hot water.
7. Moisten the fluted paper, held in a glass funnel, with water and pour the hot contents from the flask on to the filter.
8. Wash the flask and the watch glass thoroughly three times with hot water and dry in the oven.
9. Wash the filter with hot water until the washings do not affect the colour of the blue litmus paper.
10. Put the filter paper on the clock glass or Petri dish. Dry for 1 hour in the oven at $103 \pm 2°C$ and then allow to cool.
11. Roll up the filter paper using tongs that can be rinsed or with paper cover slips on the fingers and insert it into the extraction thimble. Remove any traces of fat from the clock glass or the Petri dish, using cotton wool moistened with the extraction solvent. Place the cotton wool in the thimble.
12. Place the thimble in the extraction apparatus.
13. Wash the inside of the conical flask used in instruction 1 and the clock glass with a portion of the extraction solvent and add it to the extraction flask.
14. Pour the extraction solvent into the dried and weighed flask of the

extraction apparatus. The total solvent quantity should be one and a half to twice the capacity of the extraction tube of the apparatus.

15. Fit the flask to the extraction apparatus and reflux the solvent, on the water bath, for 4 hours.
16. After extraction, detach the flask containing the liquid from the extraction apparatus and distil off the solvent using a sand bath or water bath.
17. Evaporate the last traces of the solvent on the water bath, using a jet of air if desired.
18. Dry the extraction flask for 1 hour in the oven at $103 \pm 2°C$. Cool to room temperature in the desiccator and weigh to the nearest mg.
19. Repeat these operations until the results of two successive weighings do not differ by more than 0.1% of the mass of the test portion.
20. Verify that extraction is complete by taking a second extraction flask and extracting for a further period of 1 hour with a fresh portion of the solvent. The increase in mass should not exceed 0.1% of the mass of the test portion.

Calculation

Let: Weight (g) of the test portion $= W_1$
      Weight (g) of the extraction flask with boiling chips $= W_2$
      Weight (g) of the flask and boiling chips
      with the fat, after drying $= W_3$
Then: Total fat (%) $= [(W_3 - W_2)/W_1] \times 100$

Note
1. The fat obtained cannot be used for the determination of the characteristics of the fat.

5.3 Fat in Milk Products (Roese–Gottlieb Method)

Application

The method is applicable to milk, milk products, and ice cream.

Principle

The fat content is determined gravimetrically after extraction of the fat with diethyl ether and petroleum ether from an ammoniacal alcoholic solution of the sample.

Reagents

Aqueous ammonia, 35% w/v (sp. gr. 0.880) and 26% w/v (sp. gr. 0.908) (Note 1).

Ethanol, 95–96% v/v.

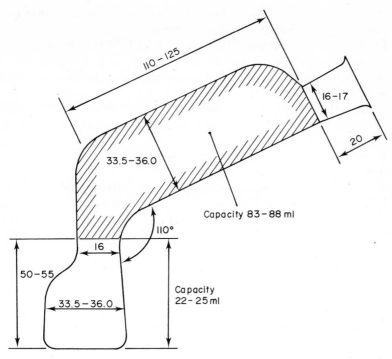

FIG. 13. Mojonnier extraction tube (dimensions in millimetres).

Diethyl ether. Peroxide free, boiling range 34–35°C.
Petroleum ether, boiling range 40–60°C.
Mixed ethers. Equal volumes of diethyl ether and petroleum ether.

Apparatus
 Air oven. Preferably fitted with a fan. Temperature 100 ± 2°C.
 Extraction tubes. Mojonnier tubes, with ground glass stoppers or solid bark corks. See Fig. 13.
 Centrifuge. A Mojonnier centrifuge is optional.
 Flat-bottomed flasks. Short-necked, capacity 150 ml; weight 40–50 g.

Procedure
 1. Weigh, to the nearest mg, 4–5 g of the prepared sample into the extraction tube (Note 2).
 2. Add 1.5 ml of aqueous ammonia 35% v/v and mix well (Note 1).
 3. Add 7 ml of warm water and again mix well (Note 3).
 4. Warm to 60–70°C and maintain at this temperature for 15 minutes.
 5. Add 10 ml of ethanol, mix, and cool (Note 4).

6. Add 25 ml of diethyl ether, close the tube with the cork or stopper (see Note 5), and shake vigorously for 1 minute.
7. Cool, remove the stopper, and, with 25 ml of light petroleum, wash the stopper and neck of the tube so that the washings run into the tube.
8. Replace the stopper, again wetted with water, and shake vigorously for 30 seconds (Note 6).
9. Support the tube on the flat bottom of the lower bulb for 30 minutes or until the ethereal layer is clear and completely separated from the aqueous layer (Note 7).
10. Remove the stopper and rinse it with mixed solvent into the extraction tube.
11. Raise the interface between the two layers to the narrowest part of the tube (if necessary) by carefully adding a little water down the side of the tube.
12. Decant carefully as much as possible of the ethereal layer into a 150 ml flask. Add 10 ml of mixed solvent to the tube and, without shaking, transfer the solvent to the flask.
13. Rinse the outside neck of the tube with mixed solvent and collect the rinsings in the flask.
14. Remove the solvents from the flask by distillation.
15. Repeat the extraction and decantation procedure twice, adding successively 5 ml of ethanol, 25 ml of diethyl ether, and 25 ml of light petroleum for each extraction.
16. Distil all remaining solvents from the flask.
17. Dry the residual fat in an oven at $100 \pm 2°C$ for 1 hour.
18. Place the flask in a desiccator to cool for at least 30 minutes, and weigh.
19. Repeat instructions 17 and 18 to constant weight.
20. Extract the fat from the flask completely by repeated washing with light petroleum, allowing any sediment to settle before each decantation.
21. Dry the residue in an oven at $100 \pm 2°C$ for 1 hour.
22. Place the flask in a desiccator to cool for at least 30 minutes, and weigh.
23. Perform a blank determination with water in place of the sample and using the specified quantities of reagents throughout.

Calculation

Let: Weight (g) of sample $= W_1$
       Weight (g) of flask + extracted fat $= W_2$
       Weight (g) of flask after removal of the fat $= W_3$
       Weight (g) of extractable material found in blank $= W_4$

Then: Fat (%) $= \dfrac{W_2 - (W_3 + W_4)}{W_1} \times 100$

Notes
1. An alternative procedure for instructions 3 and 4 is:
    3. Add 2 ml of aqueous ammonia (26 % v/v) and mix.
    4. Add 6 ml of warm water and again mix well.
2. Weigh the sample into the extraction tube by difference.
3. The complete extraction of fat is dependent on satisfactory mixing at each stage, and it is essential that all lumps present are dispersed.
4. Before each removal of the stopper, to avoid spurting of the solvent, a slightly reduced pressure should be induced in the tube by cooling.
5. The cork or stopper must be wetted with water before each insertion and washed with solvent during each removal.
6. The pressure should be released from time to time during shaking.
7. Separation of the ethereal layer from the aqueous layer can also be achieved by centrifuging at about 1000 r.p.m. for 30 seconds.

Reference
British Standard 2472 (1966).

5.4 Total Fat (Chloroform–Methanol Extraction)

Application
The method is applicable to all types of food products when further characterisation of fat is required.

Principle
The fat is extracted from the sample by vigorous stirring with chloroform/methanol mixture at room temperature. A calculated amount of water is added which separates out two phases, the lower chloroform layer containing the fat. The chloroform layer is separated off, washed with dilute sodium chloride solution to remove extracted proteinaceous material and dried with anhydrous sodium sulphate. The chloroform extract is then evaporated to dryness in a tared vial and the fat residue weighed.

Reagents
Chloroform.
Methanol.
Magnesium chloride, 20 % w/v in aqueous solution.
Sodium chloride, 0.1 % w/v in aqueous solution.
Sodium sulphate.

Apparatus
Test tubes. 200 mm, stoppered, socket size 24/29.

Glass beads, 5.5–6.5 mm.
Sintered crucibles, porosity No. 1 and porosity No. 4.
Test tubes. 150 mm, stoppered, socket size 19/26.
Pasteur pipettes.
Test tubes. Graduated up to 25 ml, stoppered.
Weighing vials (or 50 ml flasks).
Whirlimixer, For swirling test tubes.
Centrifuge.

Procedure
1. Weigh 5 g of the sample into a 200 mm long test tube.
2. Add six medium sized glass beads.
3. Add 5 ml of chloroform.
4. Add 10 ml of methanol.
5. Add 0.05 ml of 20% magnesium chloride solution (this is to reduce emulsion formation).
6. Mix on a Whirlimixer for 2 minutes.
7. Add a further 5 ml of chloroform and mix again for 2 minutes.
8. Add distilled water to bring the total water content (including that of the sample) to 9.0 ml. For example, if sample is 70% water add 5.5 ml water to the tube.
9. Mix for an additional half a minute.
10. Filter the extract through a sintered crucible porosity No. 1 (Note 1) using gentle suction, into a 150 mm long test tube.
11. Wash the extraction tube and residue on the sinter with 3×2.5 ml chloroform.
12. Swirl the tube contents and then centrifuge at 1500 r.p.m. for 5 minutes.
13. With a Pasteur pipette attached to a suction line carefully remove as much of the top aqueous layer as possible without disturbing the chloroform layer.
14. Add 10 ml 0.1% sodium chloride solution to the chloroform extract and mix by gentle inversion at least half a dozen times. (Release pressure after each inversion). Too vigorous shaking will form an emulsion (Note 2).
15. Centrifuge at 1500 r.p.m. for 5 minutes.
16. Remove the top aqueous layer as in instruction 13.
17. Add 1–2 g anhydrous powdered sodium sulphate, stopper the tube tightly and shake vigorously to dry the chloroform.
18. Filter through a sintered crucible porosity No. 4 (Note 1) into a dry test tube applying gentle suction.
19. Wash the tube and crucible with 3×2.5 ml chloroform, and combine washings in the test tube.
20. Transfer the dried chloroform extract to a pre-weighed vial (or flask),

using several small portions of chloroform to wash out the test tube.
21. Place the vial on a steam bath and evaporate off the chloroform.
22. When all the solvent has been removed place the vial containing the fat residue in an oven at 100°C for 5 minutes.
23. Cool in a desiccator and re-weigh.

Calculation

Let: Weight (g) of sample $= W_1$
Weight (g) of vial empty $= W_2$
Weight (g) of vial + fat $= W_3$
Then: Fat (%) $= [(W_3 - W_2)/W_1] \times 100$

Notes
1. Filter paper should not be used since phospholipids are absorbed on to the paper.
2. This is to remove extracted protein.

Reference
Folch J., Lees M. and Stanley G. H. S. (1957). *J. Biol. Chem.* **226**, 497.

5.5 Fatty Acid Composition (Gas–Liquid Chromatographic Method)

Application
The method is applicable to all types of lipid containing fatty acids in the range $C_{10:0}$ to $C_{22:6}$.

Principle
The sample of fat is transmethylated and the methyl esters, after purification, separated by gas liquid chromatography.

Reagents
Transmethylation mixture. Mix 150 ml of methanol and 70 ml of toluene. Cool under a tap and add carefully, with stirring, 7.5 ml of concentrated sulphuric acid (sp. gr. 1.84).
Petroleum ether, boiling point range 40–60°C.
Sodium sulphate, granular anhydrous.
Column packing. 15% polyethylene glycol adipate on 100–200 mesh A. W. Celite.

Apparatus
Gas–liquid chromatograph. Single column Pye 104 with flame ionisation detector is suitable, set with the following conditions: carrier gas, argon 60 ml

per minute; hydrogen, 50 ml per minute; air, 400 ml per minute; oven temperature, 200°C; chart speed, 120 mm per hour. Column. Glass, length 215 cm, internal diameter 3 mm.

Microsyringe, 10 μl.

Procedure

1. Weigh 10–20 mg of fat into a 50 ml round bottomed flask (Note 1).
2. Add 10 ml of the transmethylation mixture.
3. Heat on a steam bath for $1\frac{1}{2}$ hours under reflux.
4. Cool, add 10 ml of petroleum ether and 10 ml of water.
5. Stopper the flask and shake well.
6. Allow the layers to separate and withdraw the lower aqueous layer with a pasteur pipette.
7. Repeat the washings procedure with a second 10 ml of water.
8. Add sufficient sodium sulphate (2–3 g) to clarify the ether solution.
9. Decant the clear petroleum ether layer into a tube and evaporate to dryness under nitrogen.
10. Dissolve the residue in a small volume of petroleum ether (0.3 ml).
11. Inject a 5 μl aliquot of the fatty acid methyl ester solution on to the column.
12. Increase the amplifier sensitivity after the elution of linoleic acid (C 18:2) if long-chain fatty acids greater than C 20:0 are present.
13. Identify the eluted methyl esters by reference to known esters and by determining their retention volumes.

Calculation

Chromatograms of the separated fatty acid methyl esters are obtained on the potentiometric recorder. Peak areas may be measured manually or using an electronic integrating device connected to the detector output. Manual peak area measurement is conveniently obtained from peak height and peak width at 0.607 of peak height. (Make appropriate corrections if the attenuator has been changed during the analysis.)

Assume the response of the flame ionisation detector to fatty acid methyl esters to be constant; then peak area is proportional to weight of fatty acid.

Let: Peak area of individual fatty acids (A, B, C etc.) $= a, b, c, \ldots$
Then: Fatty acid content (%) of A $= [a/(a+b+c)] \times 100$
       Similarly for fatty acids, B, C, etc.

The identification of the fatty acids $C_{10:0}$ to $C_{18:2}$ usually presents no problem. However, the number of possible isomers of the C_{20} fatty acids makes assignment more difficult. There are three ways of assigning fatty acid methyl esters.

(a) Graphical plot of $\log_{10}$ retention time against carbon chainlength. Using this technique the saturated fatty acids are given whole carbon numbers (e.g., palmitic, $C_{16:0}$ has a carbon number of 16.0).

Thus, from a mixture of known fatty acid methyl esters, the carbon numbers can be determined for a series of standards and related to unknown peaks from a sample.

(b) Analysis using a non-polar stationary phase. The fatty acid methyl esters are eluted from a polar column such as polyethylene glycol adipate in order of increasing polarity for any series with the same number of carbon atoms (i.e., the retention volume varies $C_{18:2} > C_{18:1} > C_{18:0}$). This order of elution is reversed when using a non-polar column (e.g., 10% Apiezon L on 100/120 mesh A. W. Celite). In this way the retention volumes of the unsaturated fatty acids can be appreciably reduced. Thus separate analyses using polar and non-polar systems will result in two sets of carbon numbers which may be used for assignment purposes. The identification of fatty acid methyl esters from retention data as described is more certain when both a polar and non-polar phase are used.

(c) Gas chromatography/mass spectrometry. These two techniques are combined when further confirmation is required.

Notes
1. Fat isolated by Method 5.4 is preferred.

Ash, Elements, and Inorganic Constituents

6.1 Ash

Application

The method is applicable to all types of food products, with the exception of high fat (>50%) foods.

Principle

Organic matter is burned off at as low a temperature as possible and the inorganic material remaining is cooled and weighed. Heating is carried out in stages, first to drive off the water, then to char the product thoroughly and finally to ash at 550°C in a muffle furnace.

Apparatus

Muffle furnace. Thermostatically controlled at 550°C.
Electric hotplate. With thermostatic control.
Silica dishes. Diameter 8 cm, depth 2.5 cm (Note 1).
Desiccators. With fresh silica gel desiccant.

Procedure

1. Place the requisite number of silica dishes into a muffle furnace for 15 minutes or more.
2. Remove the dishes, cool in a desiccator for at least 1 hour and, when cool to room temperature, weigh each dish to the nearest mg.
3. Accurately weigh, to the nearest mg, approximately 5 g of material into each dish.
4. If the sample is a liquid, pre-dry on a steam bath to prevent spitting during the charring stage.
5. Place the dishes on a hot plate under a fume-hood and slowly increase the

temperature (Note 2) until smoking ceases and the samples become thoroughly charred.
6. Place the dishes inside the muffle furnace, as near to the centre as possible and ash overnight at 550°C.
7. Remove the dishes from the muffle and place in a desiccator for at least 1 hour to allow to cool. (The ash should be clean and white in appearance. If traces of carbon are still evident, cool the dish, add a few ml of water and stir with a glass rod to break up the ash. Dry on a steam bath and then return to the muffle furnace for 24 hours.)
8. When cool to room temperature reweigh each dish + ash to the nearest mg.
9. By difference, calculate the weight of ash.

Calculation

Let: Weight (g) of sample $= W_1$
       Weight (g) of ash $= W_2$
Then: Ash value (%) $= (W_2/W_1) \times 100$

Notes
1. If the ash is used for trace element analyses or P determination clean the dishes by boiling in 6N hydrochloric acid and rinse with water.
2. Too rapid heating is to be avoided since some of the salts will fuse and adsorb carbon which is then difficult to ignite. The use of too high a temperature may also cause some losses of volatile salts such as sodium and iron chloride.

6.2 Wet Digestion of Food Products for Element Analyses

Application
The method can be applied to all types of food products.

Principle
Organic matter is destructed and oxidized by the action of boiling sulphuric acid and nitric acid.

Reagents
Sulphuric acid, concentrated (sp. gr. 1.84).
Nitric acid, concentrated (sp. gr. 1.42).
Acid washing liquid for glassware. Dilute concentrated nitric acid 1 to 100 with water.

Check the acids for their metal content. Use reagents that contain no more than 1 mg of metal to be determined per litre.

Water. Use good quality distilled or dimineralised water.

Apparatus

Wash all glassware with acid washing liquid. Keep glassware for trace metal determination apart from other glassware.

Procedure

1. Weigh 2 g (if sample contains <10% water) or 5 g (if sample contains >10% water) of the homogenised sample into a 100 ml Kjeldahl flask.
2. Add 10 ml of concentrated sulphuric acid and shake vigorously, ensuring that no dry lumps remain.
3. Add 5 ml of concentrated nitric acid and mix.
4. Heat cautiously until the initial vigorous reaction has subsided (Note 1).
5. Heat more strongly until most of the nitrous fumes are removed.
6. Continue dropwise addition of nitric acid, until all organic matter is destroyed.
7. Heat until white fumes of sulphuric are evolved.
8. Prepare a blank using the same amounts of reagents.
9. For measurement of the element content follow the instruction given for the relevant metal.

Note

1. All operations should be carried out in a fume cupboard with proper ventilation and washing of exhaust fumes.

6.3 Metals (Atomic Absorption Method)

Application

The determination of calcium, copper, iron, magnesium, manganese, potassium, sodium, and zinc in food products.

Principle

After removal of organic material by dry ashing or wet digestion the residue is dissolved in dilute acid. The solution is sprayed into the flame of an atomic absorption apparatus and the absorption or emission of the metal to be analysed is measured at a specific wavelength.

Reagents

Hydrochloric acid, 6N, 3N, and 0.3N.

TABLE 5

Metal	Reagent	Wt reagent (g) per 250 ml of solution
Calcium	$CaCO_3$ (dry)	0.624
Copper	$CuSO_4.5H_2O$	0.981
Iron	$Fe_2 (SO_4)_3 (NH_4)_2 SO_4.24H_2O$	2.158
Magnesium	$MgSO_4.7H_2O$	2.530
Manganese	$MnSO_4.4H_2O$	1.015
Potassium	KCl (dry 2 hours at 105°C)	0.476
Sodium	NaCl (dry 2 hours at 105°C)	0.636
Zinc	$ZnSO_4.7H_2O$	1.100

Lanthanum chloride, 10% w/v.

Distilled water. Good quality or deionized water.

Filter paper. Whatman 541 or Schleicher and Schüll No. 589–1. Wash the filter papers before use with 3N hydrochloric acid to remove trace of metals.

Stock standard solutions, 1000 mg/l. Weigh out the quantities of A.R. reagents given in Table 5. Dissolve the salts in 25 ml of 3N hydrochloric acid and dilute to 250 ml with water.

Standard solutions. Dilute the stock standard solution with water (if wet digestion is applied) or 0.3N hydrochloric acid (if dry ashing is applied) to concentrations that fall within the working range. Add other salts if necessary as indicated in the first column of Table 6.

Reference standards are supplied by J. T. Baker Chemical Ltd. and Merck Chemicals, Darmstadt. Both firms sell solutions in ampoules containing an exact unit weight of an ionic species. These solutions can be diluted to a definite volume.

Apparatus

Atomic absorption apparatus. The instrument requires calibration with known standards for each series of determinations on each element.

Special glassware for trace metal analysis. All glassware must be thoroughly washed with dilute nitric acid before use. Keep glassware used for trace metal analysis apart from other glassware.

Procedure

(a) *Solutions obtained from wet digestion (Method 6.2)*:
 1. Transfer the contents of the digestion tube to a graduated flask (Note 1).
 2. Make up to the mark with water and mix thoroughly.

TABLE 6. Recommended condition for metal analyses
Flame system = air acetylene

Element	Wavelength	Absorption (A) or Emission (E)	Limit of detection* (μg metal/ml)	Working range* (μg metal/ml)
Calcium				
(+0.5% LaCl)	422.7	A	0.01	0.05–5
Copper	324.8	A	0.005	0.05–5
Iron	248.3	A	0.03	0.05–5
Magnesium	285.2	A	0.001	0.02–2 (0.5)†
Manganese	279.5	A	0.005	0.2–5 (3)
Potassium				
(+1000 μg Na/ml)	766.5	A	0.002	0.1–5 (2)
Potassium				
(+1000 μg Na/ml)	766.5	E	0.002	1–20
Sodium				
(+1000 μg K/ml)	589.0	A	0.002	0.1–5 (1)
Sodium				
(+1000 μg K/ml)	589.0	E	0.002	1–20
Zinc	213.9	A	0.004	0.1–2 (1)

*With 10-fold scale expansion. These figures are only indicative, since they depend on the apparatus and conditions.

†Values between brackets indicate that the calibration curve is linear up to that value.

(b) *Ash obtained from dry ashing (Method 6.1):*

1. Treat the ash with 5–10 ml of 6N hydrochloric acid to wet it completely, and carefully take to dryness on a low temperature hot plate.
2. Add 15 ml of 3N hydrochloric acid and heat the dish on the hot plate until the solution just boils.
3. Cool and filter through a filter paper into a graduated flask (Note 1) retaining as much of the solids as possible in the dish.
4. Add 10 ml of 3N hydrochloric acid to the dish and heat until the solution just boils.
5. Cool and filter into the graduated flask.
6. Wash the dish at least three times with water; filter the washings into the flask.
7. Wash the filter paper thoroughly and collect washings in the flask.
8. If calcium is to be determined add 5 ml of lanthanum chloride solution per 100 ml of solution.
9. Cool and dilute the contents of the flask to the mark with water.
10. Prepare a blank by taking the same amount of reagents through instructions 1–9.

(c) *Calibration of the apparatus and measurement of the samples*:
1. Set the apparatus according to the instructions.
2. Measure the calibration solutions and the reagent blank solution.
3. While running the samples, periodically check that the calibration values remain constant.
4. For the metals required prepare a calibration curve by plotting the absorption or emission values against the metal concentration in $\mu g/ml$.

Calculation

Read from the graph the metal concentration ($\mu g/ml$) that correspond to the absorption or emission values of the samples and blank.

Let: Weight (g) of samples $= W$
      Volume (ml) of extract $= V$
      Concentration ($\mu g/ml$) of sample solution $= a$
      Concentration ($\mu g/ml$) of blank solution $= b$
Then: Metal content (mg/100 g) $= [(a-b) \times V]/10W$
      or (mg/1000 g) $= [(a-b) \times V]/W$

Note
1. Select a graduated flask with suitable capacity to give metal concentrations in final solution within the working range.

6.4 Potassium and Sodium (Flame Photometric Method)

Application

This method is applicable to soups and meals having a low sodium and potassium content (i.e. dietetic foods).

Principle

After wet digestion the sodium or potassium content is measured flame photometrically.

Reagents

Nitric acid. Dilute nitric acid (sp. gr. 1.42) 1:1 with water.

Filter paper. Diamter 9 and 12.5 cm, Schleicher and Schüll No. 589–1.

Acid washing liquid for glassware. Dilute nitric acid (sp. gr. 1.42) to 100 ml with distilled water.

Distilled water. Good quality distilled or deionised water.

Sodium stock solution, 100 mg Na/l. Dry some sodium chloride at 105°C for 2 hours. Weigh 254.2 mg of the dried product, dissolve it in water, and make up to 1000 ml. Keep this solution in a polyethylene bottle.

Potassium stock standard, 100 mg K/l. Dry some potassium chloride at 105°C for 2 hours. Weigh 190.7 mg of the dried product in water and make up to 1000 ml. Keep this solution in a polyethylene bottle.

Potassium diluting solution. Weigh 3.8 g of potassium chloride, dissolve in water, and dilute to 1 litre.

Sodium working standard solutions. Pipette 0, 5, 10, 15, 20 ml of sodium stock solution into 100 ml graduated flasks. Add 5 ml of potassium diluting solution to each and dilute to the mark with water. These solutions contain 0, 0.5, 1.0, 1.5, and 2.0 mg of sodium per 100 ml respectively.

Potassium working standard solutions. Pipette 0, 5, 10, 15, 20 ml of potassium stock solution into 100 ml graduated flasks and dilute to the mark with water. These solutions contain 0, 0.5, 1.0, 1.5, 2.0 mg of potassium per 100 ml respectively.

Apparatus

Flame filter photometer or flame emission spectrophotometer.

Special glassware for trace metal analysis. Wash all glassware before use thoroughly with diluted nitric acid and after that with good quality distilled or deionised water.

Procedure

(a) *Preparation of the sample*:
1. Weigh to the nearest mg, 2 g of the sample on a filter paper (diameter 9 cm).
2. Fold up the filter paper and transfer into a 250 ml Kjeldahl flask.
3. Add 20 ml of 1:1 dilute nitric acid.
4. Boil gently for about 10 minutes and cool to room temperature.
5. Filter the digested solution through a filter paper (diameter (12.5 cm) into a 100 ml graduated flask. Wash the Kjeldahl flask and the filter paper three times each with 10 ml of water.
6. Make up to 100 ml and mix (solution A).
7. Prepare a blank starting from instruction a.2 (solution B).

(b) *Dilution of the solution for sodium*:
1. Pipette 50 ml of solutions A and B and 5 ml of potassium dilution solution into 100 ml graduated flasks, make up to the mark and mix (solution C and D respectively, see note).

(c) *Dilution of the solution for potassium determination*:
1. Pipette 5 ml of solution A and B into a 100 ml graduated flask, make up to the mark, and mix (solution E and F).

(d) *Measurement*:
1. Measure the standard solutions for both sodium and potassium. Use the right filters and in case of a spectrophotometer the wavelengths 589.0 nm and 766.5 nm.
2. Correct the obtained values for the zero concentration standards.
3. Measure the solutions C and D for sodium and solution E and F for potassium.

Calculation
   Prepare a calibration graph for sodium and potassium.
   Read the concentrations (mg/100 ml) of solutions C, D, E, and F from the calibration graph.

Let: The concentrations (mg/100 ml) be c, d, e, and f respectively.
        Weight (g) of the sample $= W$
Then: Sodium content (mg/100 g product) $= [(c-d) \times 200]/W$
        Potassium content (mg/100 g product $= [(e-f) \times 2000]/W$

Note
   Without addition of potassium to the sodium standard the results from sodium may be too high. This is caused by the inter-element effect of potassium present in the product on sodium.
   Additions of a surplus of potassium to standard and sample level out this effect. The effect depends on the temperature of the flame. An acetylene–air flame gives a greater deviation than a natural gas–air flame.

6.5 Chloride (Rapid Volhard Method)

Application
   The method can be applied to product components and meals.

Principle
   Chlorides are precipitated by an excess of silver nitrate and the organic matter is then oxidised by potassium permanganate in acid solution. The excess of permanganate is decomposed by sucrose. The unused silver nitrate is determined by titration with thiocyanate. The sodium chloride content is calculated from the amount of silver nitrate used.

Reagents
   Nitric acid reagent. Dilute concentrated nitric acid (sp. gr. 1.42) with one quarter of its volume of water.

Silver nitrate solution, 0.05N.
Ammonium or potassium thiocyanate solution, 0.05N, standardised.
Ferric ammonium sulphate solution, 5% w/v solution in 10% nitric acid.
Potassium permanganate, saturated aqueous solution.
Acetone.
Sucrose.
Urea.

Procedure
1. Weigh a suitable quantity of sample, containing 30–50 mg of sodium chloride.
2. Transfer it to a 250 ml conical flask and add 30 ml of distilled water and 2 glass beads.
3. Warm the flask and contents and add by pipette 25 ml of silver nitrate solution.
4. Add 10 ml of nitric acid reagent.
5. Boil gently in a fume cupboard for 5 minutes.
6. Cool to about 80°C and cautiously add a slight excess of permanganate to the solution.
7. Boil gently until the brown or pink colour disappears.
8. Add 0.5 ml of permanganate and boil again.
9. Repeat this treatment until the solution is not decolorised after 5 minutes boiling.
10. Add small amounts of sucrose and boil until the solution is colourless.
11. Cool for 30 seconds. Add approximately 0.1 g of urea and cool for 10 minutes in running water.
12. Add 5 ml of acetone and 2 ml of ferric ammonium sulphate indicator.
13. Titrate with the thiocyanate solution to a pink end-point.
14. Carry out a blank determination omitting the sample.

Calculation

Let: Weight (g) of sample taken $= W$
      Volume (ml) of thiocyanate used $= V_1$
      Volume (ml) of thiocyanate used in blank $= V_2$
      Normality of thiocyanate $= N$

Then: Sodium chloride (mg/100 g) $= \dfrac{(V_2 - V_1) \times N \times 5850}{W}$

6.6 Chloride (Potentiometric Method)

Application
    The method can be applied to product components and meals and is especially suitable for the determination of chloride in dietetic food products.

Principle
The salt is extracted into aqueous ethanol and the chloride determined potentiometrically with silver nitrate.

Reagents
Nitric acid, 2.0N.
Acidified ethanol, 5% v/v concentrated nitric acid (sp. gr. 1.42) in ethanol.
Ethanol, 96%.
Silver nitrate, 0.01N, standardised.

Apparatus
Recording potentiometer. Methrohm E 436 or other suitable apparatus.
Silver chloride indicator electrode.
Mercury/mercuric sulphate reference electrode.

Procedure
1. Weigh 10 g of sample (±0.1 g) into a 200 ml beaker.
2. Add 25 ml of water and boil the slurry for 1 minute.
3. Cool to room temperature and add 45 ml of ethanol. Mix and let stand for about 15 minutes.
4. Filter the slurry and wash three times with 5 ml of acidified ethanol.
5. Transfer the filtrate quantitatively with ethanol into a 100 ml graduated flask, and make up to the mark with ethanol.
6. Pipette 25 ml into a 100 ml beaker, add 0.5 ml of 2N nitric acid, and titrate potentiometrically with standard silver nitrate solution.

Calculation

Let: Weight (g) of sample $= W$
              Volume (ml) silver nitrate solution $= V$
              Normality silver nitrate solution $= N$
Then: Chloride content (mg/100 g) $= (14200 \times V \times N)/W$

6.7 Ammonia (Colorimetric Method)

Application
The method can be applied to dietetic food components and meals.

Principle
Ammonia is extracted using mild alkaline aqueous ethanol. The extract is treated with magnesium oxide and ammonia distilled into sulphuric acid. Ammonia is then determined colorimetrically after reaction with hypochlorite and phenol.

Reagents

Hydrochloric acid, 2N.

Sulphuric acid, 0.01N.

Barium hydroxide. Saturated solution at room temperature.

Hypochlorite reagent. Dissolve 12.5 g of sodium hydroxide and 3.0 ml of 5M sodium hypochlorite solution in 100 ml water. Dilute to 500 ml in a graduated flask. Store in a refrigerator.

Magnesium oxide.

Phenol reagent. Dissolve 25 g of phenol and 125 mg of sodium nitroprusside in 100 ml water. Dilute to 500 ml in a graduated flask. Store in a refrigerator.

Barium chloride, 10 % w/v.

Ethanol, 96 % w/v.

Ammonium sulphate. Dried at 105°C for 16 hours.

Ammonium sulphate stock solution. Weigh 0.388g of dried ammonium sulphate. Transfer to a 1000 ml graduated flask with water and dilute to the mark.

Ammonium standard solution. Dilute the stock solution 10-fold. Prepare the standard solution by pipetting 0, 5, 10, 15 and 20 ml of the diluted stock solution, add to each 10 ml sulphuric acid 0.01N and make up to the mark in a 100 ml graduated flask. These solutions contain 0, 50, 100, 150 and 200 μg NH_3 per 100 ml (Note 1).

Indicator paper.

Apparatus

Spectrophotometer. For the visible region or a suitable colorimeter.

Glass cuvettes. Optical pathway 1 cm.

Distillation apparatus (see Fig. 14).

Procedure

(*a*) *Extraction*:

1. Weigh 3 g (± 2 mg) of sample into a 250 ml beaker.
2. Add 75 ml of water; stir until a good suspension is obtained.
3. Add 30 ml of barium chloride solution and then cool in a refrigerator.
4. Make the suspension just alkaline with the barium hydroxide reagent, and then add 30 ml excess.
5. Add 50 ml of alcohol while stirring, and make up to the mark in a 250 ml graduated flask and mix.
6. Filter through a filter paper and pipette 75 ml of the filtrate into a 100 ml graduated flask.
7. Neutralise with hydrochloric acid to pH 7, and make up to the mark. Use indicator paper (solution 1).

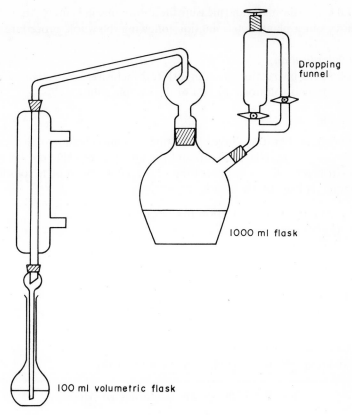

Dropping funnel

1000 ml flask

100 ml volumetric flask

FIG. 14. Distillation apparatus for ammonia determination.

(b) *Distillation*:
1. Weigh 5 g of magnesium oxide into the flask and add 350 ml of water.
2. Distil 100 ml to free the water from traces of ammonia.
3. Pipette 50 ml of solution 1 into the dropping funnel and add it to the flask as quickly as possible.
4. Wash the funnel with a small amount of water.
5. Distil off the ammonia into 10 ml of 0.01N sulphuric acid until the total volume is 100 ml (solution 2).

(c) *Colorimetric determination*:
1. Pipette 20 ml of solution 2 into a 50 ml graduated flask, add 10 ml of water.
2. Pipette 5 ml of the hypochlorite reagent, mix and then pipette 5 ml of the phenol reagent. Mix and leave the solution for 30 minutes to allow the colour to develop.

3. Make up to the mark, and measure the absorbance in 1 cm cuvette at 625 nm.
4. Carry out a blank determination following the whole procedure.

(*d*) *Standardisation*:
1. Follow the instruction given in (c)1–(c)3 inclusive, using 20 ml of each of the standard solutions instead of the sample solutions.

Calculation

Construct a calibration curve by plotting the concentration of the standard on the x-axis (in μg NH_3/100 ml) and the absorbance value on the y-axis.

Read from the calibration curve the ammonia concentration in solution 2 of the sample and the blank.

Let: Ammonia concentration in the sample
     (μg/100 ml) $= a$
     Ammonia concentration in the blank
     ((μg)/100 ml) $= b$
     Weight (g) of the sample $= W$
Then: The ammonia concentration
     (mg NH_3 per 100 g sample) $= (a-b)/(1.5 \times W)$

Note
1. Use good quality distilled water both for the preparation of the diluted standards and in the whole analytical procedure.

6.8 Total Phosphorus (Colorimetric Method)

Application
The method can be applied to the ash (method 6.1) of food products.

Principle
Orthophosphate in the ash extract reacts with ammonium molybdate in acid solution to form phosphomolybdic acid. This compound is reduced by ascorbic acid to give an intense blue colour which is measured colorimetrically.

Reagents
Sulphuric acid, 5N. Dilute 140 ml of sulphuric acid (sp. gr. 1.84) to 1 litre.
Hydrochloric acid, 3N and 6N.
Sodium hydroxide, 0.3N. Dissolve 12 g of sodium hydroxide in 1 litre water.
Stock colour reagents. Dissolve 6 g of ammonium molybdate and 0.137 g of potassium antimonyl tartrate ($KSbC_4H_4O_7.\frac{1}{2}H_2O$) in 400 ml of water. Add 500 ml of 5N sulphuric acid and mix. Dilute to 1 litre with water and remix. Store.

Working colour reagent. Add 0.53 g of ascorbic acid to each 100 ml of stock colour reagent required. Prepare this reagent fresh daily.

Phosphorus stock standard solution. Dry potassium dihydrogen phosphate (KH_2PO_4) for 2 hours at 105°C. Weigh 0.286 g of potassium dihydrogen phosphate, dissolve in water, and dilute to 100 ml (2 mg PO_4^{3-}/ml).

Phosphorus working standard solution. Dilute 5 ml of stock standard solution to 500 ml with water (20 μg PO_4^{3-}/ml).

Apparatus

Automatic dispenser. To deliver 8 ml.

Spectrophotometer or colorimeter. Suitable for the 880 nm region.

Glass cuvettes. Optical pathway 1 cm.

Procedure

(a) *Ash obtained from dry ashing (Method 6.1):*
1. Treat the ash with 5–10 ml of 6N hydrochloric acid to wet it completely, and carefully take to dryness on a low temperature hot plate.
2. Add 15 ml of 3N hydrochloric acid and heat the dish on the hot plate until the solution just boils.
3. Cool and filter through a filter paper into a graduated flask (Note 1) retaining as much of the solids as possible in the dish.
4. Add 10 ml of 3N hydrochloric acid to the dish and heat until the solution just boils.
5. Cool and filter into a 250 ml graduated flask.
6. Wash the dish at least three times with water, filter the washings into the flask.
7. Wash the filter paper thoroughly and collect the washings in the flask.
8. Cool and dilute the contents of the flask to the mark with water.

(b) *Determination of phosphorus:*
1. Pipette an aliquot from the 250 ml ash extract into a 50 ml graduated flask. This aliquot must contain less than 100 μg of phosphate and more than 20 μg of phosphate.
2. Add an equal volume of 0.3N sodium hydroxide to neutralise the solution.
3. Add water to make the total volume approximately 30 ml.
4. Using the automatic dispenser add 8 ml of working colour solution.
5. Dilute to volume with water and mix.
6. Allow colour to develop for 10 minutes (Note 1).
7. Read the absorbance in a 1 cm cuvette at 882 nm.

(2) *Standardisation:*
1. Pipette 0, 1, 2, 3, 4, and 5 ml aliquots of working standard solution into

50 ml graduated flasks. These solutions contain 0, 20, 40, 60, 80, and 100 μg PO_4^{3-} per 50 ml.
2. Add water to make volume approximately 30 ml.
3. Continue as in instructions 1.4 to 1.7.

Calculation

Construct a calibration curve by plotting the concentration of the standards on the x-axis (g as PO_4^{3-} per 50 ml) and the absorbance value on the y-axis.

Read from the calibration curve the phosphate concentration of the sample and blank.

Let: Phosphate (μg as PO_4^{3-}/50 ml) for the sample $= a$
      Phosphate (μg as PO_4^{3-}/50 ml) for the blank $= b$
      Weight (g) of sample taken for ash analysis $= W$
      Aliquot (ml) taken for analysis $= v$

Then: Phosphate content ($\% PO_4^{3-}$) $= \dfrac{0.025 \times (a-b)}{W \times v}$

or

Phosphorus content ($\% P$) $= \dfrac{0.00815 \times (a-b)}{W \times v}$

Note
1. The colour is stable for several hours.

6.9 Total Phosphorus (Colorimetric Method, Automated)

Application

The method can be applied to all types of foodstuffs.

Principle

The sample is wet digested according to Method 6.2 during which all the phosphorus is converted to orthophosphoric acid. Using an automatic analysis system an aliquot of the digest is reacted with ammonium molybdate and hydrazine in acid solution to yield a blue colour which is measured at 660 nm.

Reagents

Sulphuric acid (3.1N). Dilute 86 ml of concentrated sulphuric acid (sp. gr. 1.84) to 1 litre with water.

Sulphuric acid (1.0N). Dilute 28 ml of concentrated sulphuric acid (sp. gr. 1.84) to 1 litre with water.

Sodium hydroxide solution (0.2N). Dissolve 8 g of sodium hydroxide in water and dilute to 1 litre.

Ammonium molybdate solution (2.5 % w/v). Dissolve 6.25 g of ammonium molybdate [(NH$_4$)$_6$ Mo$_7$O$_{24}$ 4H$_2$O] in 200 ml of water and make up to 250 ml.

Hydrazine solution (0.26 % v/v). Mix 4.3 ml of 24 % hydrazine hydrate with 7.5 ml of 1N sulphuric acid and make up to 250 ml with water.

Phosphorus stock standard solution. Dry potassium dihydrogen phosphate (KH$_2$PO$_4$) for 2 hours at 105°C. Weigh 176 mg and make up to 500 ml with 3.1N sulphuric acid. 1 ml of this standard corresponds to 0.080 mg P.

Phosphorus working standard solution. Pipette 1, 3, 5, and 10 ml aliquots of stock standard into 100 ml graduated flasks and dilute to the mark with 3.1N sulphuric acid. These solutions contain 0.08, 0.24, 0.40, and 0.80 mg per 100 ml respectively.

Apparatus

Continuous flow analyser. For automatic analysis with appropriate accessories (see Fig. 15).

Procedure

1. Weigh an amount of sample corresponding to 500 mg of dry matter into a 100 ml Kjeldahl flask.
2. Digest the sample as described in Method 6.2.
3. Transfer the digest to a 100 ml graduated flask and dilute to the mark with 3.1N sulphuric acid.
4. Pipette an aliquot corresponding to about 0.5 mg of P into a 100 ml graduated flask and dilute to the mark with 3.1N sulphuric acid (Note 1).

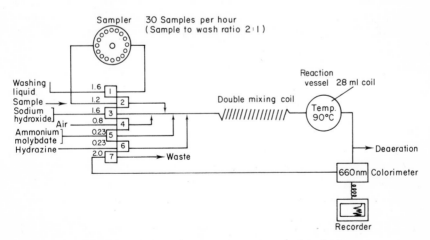

FIG. 15. Set-up for determination of total phosphorus.

5. Assemble the apparatus as in Fig. 15. Pump the reagents through until the system has reached equilibrium.
6. Fill the sampler cups with working solutions followed by sample solutions. The range of standards should be repeated after every 12–15 samples.
7. Prepare blanks starting from Step 3.
8. Run the system until all samples, standards, and blanks have passed the flow cell of the colorimeter.

Calculation

Read the absorbance values of blanks, samples, and standard solutions. Correct the values of standards and samples for the blank value. Prepare a calibration curve with on the x-axis the phosphate concentration expressed as mg phosphorus per 100 ml solution (0.08–0.24–0.40–0.80 mg respectively) and on the y-axis the absorbance values.

Read the P-values (mg/100 ml) of the sample from the calibration curve.

Let: The phosphorus content of solution
      (mg P per 100 ml) $= a$
      Weight (g) of sample $= W$
      Dilution factor $= F$
Then: Total phosphate content (as % PO_4^{3-}) $= (a \times F \times 0.306)/W$
      or Total phosphorus content (as % P) $= (a \times F)/10 \times W)$

Notes

1. The following dilutions with 3.1N sulphuric acid are recommended:
   P content (%)   dilution factor
   0 to 0.2   1 (no dilution)
   0.2 to 1   5
   1 to 5   20

Fat-Soluble Vitamins

7.1 Vitamin A and/or β-Carotene (Manual Method)

Application

The method is applicable to fats and most other foods.

Principle

The sample is saponified and the unsaponifiables extracted with diethyl ether. Vitamin A is separated from carotenoids by alumina chromatography, and β-carotene from other carotenes by magnesia chromatography. Both are then determined spectrophotometrically.

Reagents

Potassium hydroxide solution (60% w/w aqueous). Dissolve 160 g of potassium hydroxide in 100 ml of water.

Sodium hydroxide solution. Dissolve 10 g of sodium hydroxide in water and dilute to 100 ml.

Carr Price reagent. Add 200 g of antimony trichloride to 800 ml of chloroform in a 2 litre flask. Heat to dissolve under reflux. Cool, add 30 ml of acetic anhydride, and dilute to 1 litre with chloroform. Store in a dark bottle.

Sodium sulphate. Granular, anhydrous. Dry in an oven at 100°C for 3 hours before use.

Ethanol. Absolute.

Petroleum ether. Boiling range 40–60°C, dried over anhydrous sodium sulphate.

Diethyl ether. Peroxide-free, dried over anhydrous sodium sulphate.

Elution solutions. 4, 8, 12, 16, 20, 24, 36, and 50 per cent dry diethyl ether in dry petroleum ether.

Quinol.

Alumina. Prepare from alumina trihydrate (Note 1).

Neutral alumina. Prepare this alumina as follows:
1. Activate the fraction which passes through 150 mesh sieve (British Standard) at 800°C for 7 hours.
2. Cool.
3. Add 2 g of water per 98 g of activated alumina and mix well.
4. Store in a stoppered bottle.
5. Check that, using this alumina, the vitamin A starts to come off the column at the 24% solvent stage (see chromatographic separation).
6. If the vitamin A comes off too late, add more water to the alumina until the right conditions are obtained.

Alkaline alumina. Prepare this alumina as follows:
1. Weigh 20 g of neutral alumina into a 100 ml round-bottom stoppered flask.
2. Add 20 ml of 10% sodium hydroxide solution.
3. Attach an adapter, with side arm, to the flask.
4. Stopper the adapter with a rubber bung.
5. Pass a glass tube, which has been drawn to a coarse capillary, through the rubber bung, so that the end of the tube is about 5 mm from the bottom of the flask.
6. Connect the glass tube to a hydrogen or nitrogen source.
7. Attach the side arm of the adapter to a vacuum pump.
8. Apply suction and pass a slow stream of gas through the flask for 1 hour.
9. Immerse the flask in an oil bath and raise the temperature gradually during 1 hour to 130 ± 2°C.
10. Maintain the flask at 135 ± 2°C for 1 hour.
11. Discontinue the gas stream.
12. Remove the flask from the oil bath and allow to cool under vacuum.
13. Release the vacuum with hydrogen or nitrogen.
14. Determine the moisture content of the product by drying at 500°C for 2 hours.
15. Add water to the alumina so that the final moisture content is 12.5 ± 1%.
16. Store the product in small glass tubes sealed with wax (1 g per tube).

Magnesia. Heat magnesium oxide (heavy, particle size 50 μm) at 100°C for 2 hours. Cool in a desiccator. Store in an airtight bottle for 3 days before use.

Apparatus
  Graduated tubes, 1 ml capacity.
  Pipette, 0.5 ml, the tip drawn to a capillary to fit the graduated tubes.
  Chromatographic columns. See Fig. 16.
  Spectrophotometer. Suitable for ultraviolet and visible region.
  Silica and glass cuvettes. Optical pathway 1 cm.

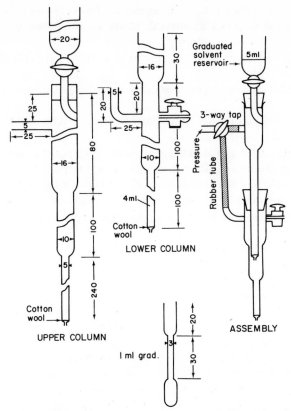

FIG. 16. Chromatographic apparatus for vitamin 'A' assay (dimensions in millimetres).

Procedure (Note 2)

(*a*) *Saponification and extraction*:
 1. Weigh an amount of sample (but not more than 25 g) containing approximately 80 µg of vitamin A, into a 250 ml flat-bottomed flask (Note 3).
 2. Add 20 mg of quinol, 60 ml of ethanol (96 % w/v), 10 ml of 60 % potassium hydroxide solution, and 10 ml of petroleum ether.
 3. Boil under reflux for 30 minutes (protect from daylight).
 4. Cool.
 5. (a) If there is no solid material left after saponification, transfer the contents of the flask with two 80 ml portions of water to a 500 ml separating flask.
      (b) If there is solid material left after saponification, filter the solution through a Büchner funnel, using a fast filter paper, into a 500 ml separating funnel. Add 160 ml of water to the extract in the funnel.

6. Add 100 ml of diethyl ether to the extract. Shake the funnel continuously, opening the stopcock at intervals to release the pressure.
7. Allow the phases to separate completely.
8. Draw off the lower aqueous phase into another separating funnel. Keep the ether layer.
9. Extract the aqueous layer three more times using 50 ml portions of diethyl ether.
10. Add 50–100 ml of water to the combined ether extracts and swirl gently.
11. Discard the lower phase.
12. Continue washing by shaking gently with 50 ml portions of water until the washings are free of alkali (test with phenolphthalein).
13. After removal of the final water wash, allow the ether extract to stand for several minutes and carefully draw off any separated water.
14. Evaporate the ether extract to dryness on a steam bath with a stream of inert gas.
15. Add 2 ml of absolute alcohol immediately all the diethyl ether has been removed (the residue must not remain dry longer than is necessary).
16. Evaporate to dryness in a current of inert gas.
17. Repeat instructions 15 and 16 once more.
18. Immediately dissolve the residue in 5 ml of petroleum ether.
19. Evaporate to dryness in a current of inert gas.
20. Repeat instructions 18 and 19 twice more.
21. Dissolve the residue in 2 ml of petroleum ether.

(b) *Determination of vitamin A – Alumina Column Chromatography*:
1. Place a small plug of cotton wool in the lower tip of the upper chromatographic column (Fig. 16).
2. Pour in petroleum ether to a level half way up the centre section of the column.
3. Add 5.5 g of neutral alumina.
4. Allow the excess petroleum ether to drain to within 2 mm of the surface of the alumina (under pressure of an inert gas).
5. Transfer the extract from (a)21 with two 1 ml portions of petroleum ether to the column.
6. Develop the column under pressure.
7. Add in sequence to the column, when the meniscus of the preceding portion approaches the surface of the alumina, 5 ml of petroleum ether then 5 ml of elution solutions in order from 4 to 20% diethyl ether in petroleum ether.
8. If there is any carotene eluted during the petroleum ether to 12% diethyl ether in petroleum ether solutions, collect and reserve for β-carotene analysis.

9. Immediately after adding the 20% diethyl ether eluant, attach the lower column containing 1 g of alkaline alumina in petroleum ether.
10. Collect the eluate in calibrated 1 ml tubes.
11. Continue developing the column with 5 ml of 24% and 36% diethyl ether solutions until all the vitamin A has been eluted.
12. Withdraw from each tube approximately 0.2 ml of solution.
13. Add approximately 0.3 ml of Carr Price reagent; a positive blue colour indicates the presence of vitamin A.
14. From each tube containing vitamin A take exactly 0.5 ml, combine in a 10 ml graduated flask, and dilute to volume with petroleum ether.
15. Measure the absorbance against a petroleum ether blank in 1 cm silica cells at 323, 324, 325, and 326 nm.
16. Also measure the absorbance at maximum wavelength -15 and maximum wavelength $+10$ (Note 4).

(c) *Determination of β-carotene; magnesia column chromatography*:
1. Evaporate the fraction containing carotene to dryness on a water bath at 50°C, with a stream of inert gas.
2. Dissolve the extract in 2 ml of petroleum ether.
3. Place a small plug of cotton wool in the lower tip of the chromatographic tube (Fig. 16, upper column).
4. Pour in petroleum ether to a level half way up the centre section of the column.
5. Add 3 g of magnesia.
6. Allow the excess light petroleum to drain to within 2 mm of the surface of the magnesia (under pressure from an inert gas).
7. Transfer the extract from (c)2 with two 1 ml portions of light petroleum to the column.
8. Develop the column under pressure.
9. Add in sequence to the column, when the meniscus of the previous portion approaches the surface of the magnesia, 5 ml of light petroleum then 5 ml of elution solutions in order from 4 to 12% diethyl ether in petroleum ether. Any α-carotene present will be eluted.
10. Add in sequence to the column the elution solutions in order from 16 to 50% diethyl ether in petroleum ether.
11. Collect the deep orange coloured zone of β-carotene.
12. Evaporate this fraction to dryness on a water bath at 50°C with a stream of inert gas.
13. Dissolve the β-carotene in petroleum ether and dilute to 25 ml.
14. Measure the absorbance against petroleum ether in a 1 cm glass cell at 2 nm intervals from 440 to 456 nm.

G*

Calculation

(a) For vitamin A

$$\begin{aligned}
\text{Let:} \quad \text{Weight (g) of sample} &= W \\
\text{Absorbance} &= D \\
E^1_{1\,cm}\% \text{ for vitamin A in petroleum ether} &= 1830 \; (\lambda \; 324 \text{ nm})
\end{aligned}$$

$$\text{Then: Vitamin A (retinol) in sample (mg/100 g)} = \frac{D \times 10^6 \times 2}{1830 \times 100 \times W}$$

$$= \frac{D \times 10.9}{W}$$

(b) For β-carotene

$$\begin{aligned}
\text{Let:} \quad \text{Weight (g) of sample} &= W \\
\text{Absorbance} &= D \\
E^1_{1\,cm}\% \text{ for } \beta\text{-carotene in petroleum ether} &= 2500 \; (\lambda \; 452 \text{ nm})
\end{aligned}$$

Then: β-carotene (retinol equivalent) in sample (mg/100 g)
[$6\mu g$ β-carotene $= 1$ μg retinol]

$$= \frac{D \times 10^6}{2500 \times 100} \times \frac{25}{W} \times \frac{100}{1000} \times \frac{1}{6} = \frac{D \times 1.67}{W}$$

Notes

1. Suitable quality can be obtained from the British Aluminium Company. Woelm Alumina (activity 1) is also suitable, if used omit steps 1–3 and instead add 10–15% water to deactivate.
2. The whole process should take place as quickly as possible in subdued light.
3. For samples such as milk the fat is first extracted by the Roese–Gottlieb method. The extracted fat is then saponified.
4. The ratios $\dfrac{\text{Abs. at}(\lambda_{max} - 15)}{\text{Abs. at } \lambda_{max}}$ and $\dfrac{\text{Abs. at}(\lambda_{max} + 10)}{\text{Abs. at } \lambda_{max}}$
(in which Abs. = Measured absorbance)
for a pure vitamin A fraction does not exceed 0.9. A ratio higher than 0.9 indicates the presence of an intefering substance.

References

Analytical Methods Committee (1964), *Analyst* **89**, 7.
The Food Standard (Margarine) Order S1 (1954) 613. *Analyst* (1968), **93**, 107.

7.2 Vitamin A and/or β-Carotene (Automated High Performance Liquid Chromatographic Method

Application

The method can be applied to all food products.

Principle

The sample is saponified and the unsaponifiable matter is extracted with hexane. Using an HPLC system, an aliquot of the hexane solution is chromatographed on alumina using 2% ethanol in benzene as eluting solution. The eluted vitamin A is determined fluorimetrically (excitation 330 nm, emission 514 nm). A second aliquot of the hexane solution is chromatographed on alumina using 0.4% dioxan in hexane as eluting solution. The eluted β-carotene is determined colorimetrically at 460 nm.

A fully automated injection and analytical system is described (see Fig. 17). Samples may also be injected manually onto the column.

Reagents

Potassium hydroxide solution, 0.5N. Dissolve 28 g of potassium hydroxide in ethanol and adjust to 1 litre.

Potassium hydroxide solution, 1N. Dissolve 56 g of potassium hydroxide in water and adjust to 1 litre.

Sodium sulphite, 10% w/v.

Sodium sulphate, anhydrous.

Eluting solution for vitamin A analysis. Mix 20 ml of absolute ethanol with 1 litre of benzene.

Eluting solution for β-carotene analysis. Mix 4 ml of dioxan with 1 litre of hexane.

Aluminium oxide. Supplier Woelm, No. 18, activity super 1, particle size 18–30 μm. Add 5 ml of water to 95 g of aluminium oxide (i.e. 5% added water) and mix thoroughly.

Vitamin A alcohol stock solution. Dissolve 10.0 mg of vitamin A alcohol in hexane and dilute to 100 ml. Stable for 1 month if stored in the dark at $-20°C$.

β-Carotene stock solution. Dissolve 10.0 mg of β-carotene in hexane and dilute to 100 ml. Stable for 1 month if stored in the dark at $-20°C$.

Vitamin A and β-carotene mixed standard solutions. Prepare fresh daily four mixed working standards (A, B, C, D) in 100 ml graduated flasks. Using a micro-syringe, add 20, 40, 60, 80 μl vitamin A stock solution and 200, 400, 600, 800 μl β-carotene stock solution respectively to each flask. Dilute to 100 ml with hexane. These solutions contain 2, 4, 6, 8 μg vitamin A and 20, 40, 60, 80 μg β-carotene per 100 ml respectively.

Apparatus

(*a*) *Saponification and extraction*:
  Glycerol bath. At 100°C.
  Round bottomed flask. 250 ml capacity, twin necked.
  Nitrogen bleed.

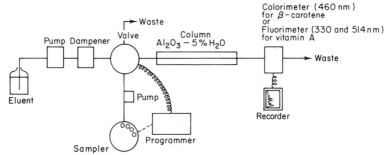
FIG. 17. Scheme of automatic assay for vitamin A or β-carotene.

Hot plate stirrer.
Condenser. Water cooled.
Separating funnel. 250 ml capacity.
Filter funnel. 200 ml capacity, with sintered glass disc.

(b) *Chromatography and determination (see Fig. 17 for schematic diagram of apparatus)*:
Glass column. 50×0.3 cm. i.d. dry packed with aluminium oxide containing 5% added water.
High pressure pump. Capable of operating up to $1000 \, lb/in^2$ and delivery up to 2.4 ml/min.
Pulse dampening system. N.B. this is not required with a pulse-free high-pressure pump.
Injection valve. With 500 μl sample loop.
Flow through colorimeter. At 460 nm, 4 cm optical path.
Flow through fluorimeter. Excitation 330 nm, emission 514 nm with internal cell volume of 100 μl.
Recorder. 10 mV full scale.
Sampler and programmer. To automate loading the column and the chromatographic determination.

Procedure (Note 1)

(a) *Saponification and extraction*:
  1. Weigh 5.0 g of the homogenised sample into a 250 ml round bottomed flask.
  2. Add 1 ml of the sulphite solution, a magnetic stirring bar, and 50 ml of 0.5N potassium hydroxide in ethanol.
  3. Attach the flask to the condenser and reflux for 30 minutes in a glycerol bath at 100°C.

4. Stir the contents of the flask during this time and flush nitrogen through the flask (about 100 ml/min).
5. Cool the content afterwards in ice and add 100 ml of hexane.
6. Homogenise by swirling and decant into the separation vessel.
7. Add 50 ml of 1N potassium hydroxide to the separator.
8. Shake for 10 seconds, allow the phases to separate for 10 minutes, and remove water layer.
9. Wash the hexane successively with 50 ml of 1N potassium hydroxide in water and 3 times with 50 ml of water.
10. Run the remaining hexane layer off into a 100 ml conical flask, add about 2 g sodium sulphate (anhydrous) and shake.

(b) *Chromatography and determination of vitamin A and/or β-carotene*:
  1. Transfer the hexane fraction to the sample tubes and cover the tubes with aluminium foil.
  2. Fill the sample tray with the tubes, reserve the first four places for standards A, B, C, and D, and further place a vitamin A/β-carotene standard after every seven tubes.
  3. Set the programmer and start the apparatus.

Calculation
  1. Measure the height of the peaks of samples and standards (Note 2).
  2. Prepare calibration curves for vitamin A and β-carotene with the height of the peaks of the standards on the y-axis and the standard concentration in microgram/100 ml on the x-axis.
  3. Read the concentration of vitamin A or β-carotene from the calibration curve.

$$\text{Let:} \quad \text{This value } (\mu g/100 \text{ ml}) \ = a$$
$$\text{Weight (g) of sample} \ = W$$

Then: Vitamin A (retinol) content (mg/100 g) in sample $= a/(W \times 10)$
β-Carotene (retinol) content (mg/100 g) in sample $= a/(W \times 60)$

Notes
  1. Laboratory conditions. During the whole procedure the samples should be protected against ultraviolet light. It is recommended that the procedure be carried out in a darkened room.
  2. Use for β-carotene recorder paper with a logarithmic scale and read the absorbances.

7.3 Vitamin D (Gas Chromatographic Method)

Application
   The method is of proven use for cod liver oil and should be applicable to other marine oils rich in vitamin D (i.e. ≥ 1 $\mu g/g$) (Note 1).

Principle

The sample, together with an internal standard, is saponified and the unsaponifiables are extracted with diethyl ether. Sterols are removed by precipitation, first from 10% aqueous methanol solution and then from 4% digitonin solution in 10% aqueous methanol, as digitonides. Vitamin D is isomerised to the isotachysterol using antimony trichloride, and the isomer is separated from vitamins A and E and other interfering substances by alumina column chromatography. The halogeno-ester of the isotachysterol is formed using heptafluorobutyric anhydride. This is extracted with petroleum ether and determined by electron-capture gas–liquid chromatography.

Apparatus

Kuderna-Danish (K.D.) evaporators.

Centrifuge.

Chromatographic column, 30 cm × 1 cm i.d. with PTFE taps with sinter.

Gas–liquid chromatograph with electron-capture detector. Pye series 104 chromatograph (or other suitable apparatus) fitted wwith nickel-63 electron capture detector.

Glass column. Length 215 cm, i.d. 3 mm, packed with 3% OV-17 on GasChrom Q 60–80 mesh.

Column temperature, 230°C.

Detector temperature, 285°C.

Nitrogen carrier gas, flow rate 65 ml/min.

Reagents

Vitamin D_2. Koch Light Laboratories Ltd., crystalline pure 4×10^7 I.U./g. Internal standard: Accurately weigh 250 mg of vitamin D_2 and dissolve in 500 ml of chloroform. Dilute 5 ml of this solution to 100 ml with petroleum ether (40–60°C) to give the working standard vitamin D_2 solution of 25 μg/ml.

Chloroform, A.R. grade.

Quinol.

Ethanol, absolute.

Potassium hydroxide solution, 60% w/v aqueous. Dissolve 60 g of potassium hydroxide in distilled water and dilute to 100 ml.

Petroleum ether, boiling point range 40–60°C.

Diethyl ether, A.R. grade.

Phenolphthalein.

Sodium sulphate, anhydrous, granular, A.R. grade.

Methanol, A.R. grade.

Digitonin, pure, from Koch Light Laboratories Ltd.

Antimony trichloride solution, 20% (w/v) in chloroform. Dissolve 20 g of antimony trichloride A. R. in chloroform; dilute to 100 ml with chloroform, and filter.

Tartaric acid solution, 40% (w/v) aqueous. Dissolve 40 g of (+)-tartaric acid A.R. in water and dilute to 100 ml.

Diethyl ether–petroleum ether. Mix equal volumes of diethyl ether and petroleum ether (boiling range 40–60°C).

Alumina. Woelm neutral activity $2\frac{1}{2}$. To 1 kg of aluminium oxide, Woelm, neutral, add 35 ml of distilled water. Shake vigorously until the exothermic reaction stops. Cool. Store in tightly stoppered jar.

Tetrahydrofuran. Puriss from Koch Light Laboratories Ltd.

Heptafluorobutyric anhydride, from Koch Light Laboratories Ltd. Using a Pasteur pipette carefully transfer the solution to small glass ampoules with rubber caps (approx. 1–2 ml per ampoule) and seal.

Sodium hydrogen carbonate solution, 2% (w/v) aqueous. Dissolve 2 g of sodium bicarbonate A.R. in water and dilute to 100 ml.

Procedure

(*a*) *Saponification and extraction:*
1. Into a 250 ml flat bottomed flask weigh between 2 g and 10 g of cod liver oil (depending on the expected potency).
2. Carefully add by pipette 1 ml of the internal standard solution of vitamin D_2.
3. Add 20 mg of quinol, 60 ml of ethanol, 10 ml of 60% potassium hydroxide solution, and 10 ml of petroleum ether.
4. Boil under reflux for 30 min (protect from daylight).
5. Cool.
6. Transfer the contents of the flask with two 80 ml portions of water to a 500 ml separating flask.
7. Add 100 ml of diethyl ether to the extract. Shake the separator gently, opening the stopcock at intervals to release the pressure.
8. Allow the phases to separate completely.
9. Draw off the lower aqueous phase into another separating funnel.
10. Extract the aqueous layer three more times using 50 ml portions of diethyl ether. Solutions at this stage may be shaken more vigorously.
11. Add 50–100 ml of water to the combined ether extract and swirl gently.
12. Discard the lower aqueous phase.
13. Continue washing by shaking gently with 50 ml portions of water until the washings are alkali-free (test with phenolphthalein).
14. After removal of the final water wash, allow the ether extract to stand for several minutes and carefully draw off any separated water.
15. Add a small amount of anhydrous sodium sulphate (granular) to the extract and shake.

16. Transfer the ether extract quantitatively to a Kuderna–Danish evaporator and evaporate until the volume has been reduced to a few millilitres.
17. Rinse the flask with a few millilitres of ether and allow to drain for a few minutes.
18. Remove the tube and evaporate the extract just to dryness under a stream of nitrogen.
19. Dissolve the residue in 9 ml of methanol (it may be necessary to warm at this stage) and add 1 ml of water. Shake the solution.

(b) *Precipitation of Sterols*:
1. Stand the solution in 10 ml of 10% aqueous methanol at 0°C for at least 1 hour.
2. Centrifuge for 5 minutes at 2000 r.p.m.
3. Decant the supernatant liquid into a centrifuge tube containing 0.4 g of digitonin dissolved in 10 ml of 10% aqueous methanol. Drain the tube carefully to transfer the maximum volume of supernatant.
4. Allow to stand overnight in the refrigerator.
5. Centrifuge for 5 minutes at 2000 r.p.m.
6. Decant the supernatant liquid into a 50 ml or 100 ml separating funnel.
7. Extract the solution by shaking gently with 10 ml of petroleum ether, releasing the pressure at intervals.
8. Allow the phases to separate completely.
9. Draw off the lower phase into a second separating funnel.
10. Transfer the petroleum ether layer to a third separating funnel containing a small amount of water.
11. Repeat the extraction twice more using 10 ml volumes of petroleum ether.
12. Wash the combined petroleum ether extracts three times with equal volumes of water, discarding the lower aqueous phase each time.
13. Dry the extract with a small amount of anhydrous sodium sulphate (granular).
14. Transfer the extract quantitatively to a 10 ml graduated tube and evaporate just to dryness under a stream of nitrogen.
15. Dissolve residue in 0.2 ml of chloroform.

(c) *Isomerisation*:
1. To the solution in 0.2 ml of chloroform add 2 ml of 20% (w/v) antimony trichloride solution in chloroform. Shake the tube and allow to stand for 15 seconds at room temperature.
2. Stop the reaction by the addition of 3 ml of 40% (w/v) tartaric acid solution in water. Shake the suspension vigorously until two distinct phases are obtained.

3. Add 5 ml of petroleum ether to the tube and shake vigorously.
4. Allow the phases to separate completely.
5. Using a Pasteur pipette transfer the petroleum ether layer to a 50 ml separator.
6. Re-extract the aqueous phase using a further 5 ml of petroleum ether.
7. Using the same Pasteur pipette transfer this petroleum ether layer to the 50 ml separator.
8. Wash the combined extracts by shaking with an equal volume of water.
9. Allow the phases to separate completely.
10. Run off the lower aqueous phase and discard.
11. Wash the extract twice more in a similar way, each time discarding the aqueous phase.
12. Add a small amount of anhydrous sodium suplhate (granular) to the extract and shake vigorously.
13. Transfer the extract to a graduated 10 ml tube and evaporate just to dryness under a stream of nitrogen.
14. Redissolve the residue in approx. 1 ml of 50% diethyl ether–petroleum ether.

(d) *Alumina column chromatography*:
1. Close the top of the chromatographic column and half fill the column with 50% diethyl ether–petroleum ether.
2. Add 12 g of neutral alumina, activity $2\frac{1}{2}$.
3. Allow the excess solvent to drain to within 2 mm of the surface of the alumina.
4. Transfer the extract [step (c)14] to the column with two 1 ml portions of 50% diethyl ether–petroleum ether, allowing the extract and rinsings to run to within 2 mm of the surface of the alumina between additions.
5. Elute the column with 50% diethyl ether–petroleum ether.
6. Discard the first 90 ml of column eluate (first fraction, Note 2).
7. Collect the next 140 ml of column eluate (second fraction, Note 2).
8. Dry the second fraction from the column with a small amount of anhydrous sodium sulphate (granular).
9. Transfer the solution to a Kuderna–Danish evaporator fitted with a 10 ml graduated tube using small quantities of 50% diethyl ether–petroleum ether to rinse the sodium sulphate.
10. Evaporate the solution to a small volume, rinse the Kuderna–Danish evaporator with small volumes of 50% diethyl ether–petroleum ether and allow to drain for several minutes.
11. Continue evaporation to dryness in the 10 ml graduated tube under a stream of nitrogen.
12. Dissolve the residue in 1 ml of tetrahydrofuran.

(e) *Esterification*:
1. Stand the solution in 1 ml of tetrahydrofuran in an ice bath.
2. Using a 250 μl syringe add 0.1 ml of heptafluorobutyric anhydride. Shake, and leave in the ice bath for 1 minute.
3. Stop the reaction by the addition of 10 ml of water.
4. Add 2 ml of petroleum ether to the solution and shake to extract the ester.
5. Allow the phases to separate completely.
6. Using a Pasteur pipette, withdraw the lower aqueous phase and discard.
7. Add 10 ml of water to the tube and shake, again allowing the phases to separate and removing the aqueous phase as before.
8. Wash the extract with 10 ml of 2% sodium bicarbonate solution, discarding the aqueous phase.
9. Again wash the extract with a further 10 ml of water. Discard the aqueous phase.
10. Allow the tube to stand for several minutes. Using a Pasteur pipette, carefully remove any water which has separated out.
11. Adjust the extract to a convenient volume with petroleum ether (usually 2 ml).

(f) *Gas liquid chromatography*:
1. Inject a suitable aliquot (usually 2 μl) of the extract on to the gas–liquid chromatographic column.
2. To determine retention times, inject an aliquot of a mixed vitamin D_2/D_3 standard which has been taken through stages c.3 and e.5 only of the procedure (approx. 5–10 ng injected gives suitable sized peaks).

Calculation

Let: Height of the vitamin D_2 peak $= H_2$
       Height of the vitamin D_3 peak $= H_3$
       Weight (μg) of vitamin D_2 added $= D_2$
       Weight (μg) of sample $= W$
       Response for vitamin D_3 is 1.5 × that for vitamin D_2.

Then: Vitamin D_3 content (μg/g) of sample
$$= \frac{H_3 \times D_2}{H_2 \times 1.5 \times W}$$

or vitamin D_3 content (I.U/g)
$$= \frac{H_3 \times D_2}{H_2 \times 1.5 \times W} \times 40$$

Notes
1. The method is not suitable in its present form for other foodstuffs containing lower levels of the vitamin, for example margarine or evaporated milk (0.01–0.10 μg/kg) owing to higher levels of interfering substances. To

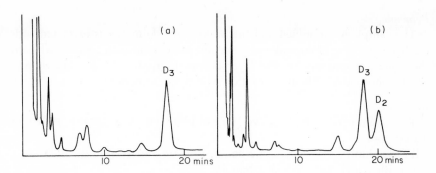

FIG. 18. Typical chromatograms from 5-gram samples of cod liver oil: (a) cod liver oil extract; (b) extract containing vitamin D_2 as internal standard.

remove these interferences, further clean-up by preparative thin-layer chromatography is required at step (e)11 and the gas chromatograph needs to be coupled with a mass spectrometer to obtain sufficient sensitivity and specificity in the determinative step.

2. Variations in the activity of the alumina have been found to occur, and it is necessary to check each batch, as prepared and at intervals after preparation. This can be achieved as follows.

2.1 Isomerise 100 μg vitamin D (see 3).

2.2 Transfer to an alumina column and elute with 50% diethyl ether–petroleum ether (see 4), collecting the following fractions: 80 ml, 5 ml, 5 ml, 120 ml, 10 ml, 10 ml, 10ml, 10ml.

2.3. Use the characteristic absorption of isotachysterol at 278, 288, and 301 nm in the various fractions, to obtain the correct elution pattern of the isomerised vitamin D.

 Alternatively, a smaller initial weight of vitamin D may be taken (10 ng) and the elution pattern determined by GLC analysis following derivatisation.

References

1. Wilson, P. W., Lawson, D. E. M., and Kodicek, E. (1966). *Biochem. J.* **98**, 293.
2. Bell, J. G. and Christie, A. A. (1973). *Analyst* **98**, 268.
3. Ueda, F. *et al.* (1971). *G. Vitaminology* **17**, 142.
4. Bell, J. G. and Christie, A. A. (1974). *Analyst* **99**, 385.

7.4 Vitamin E (Thin layer Chromatographic Method)

Application
 The method is applicable to fats and most other foods.

Principle

The sample is saponified and the unsaponifiable matter is extracted with diethyl ether. The biologically active α and β-tocopherols are separated from the other tocopherols by thin-layer chromatography and are determined spectrophotometrically after reaction with a α, α'-bipyridyl.

Reagents

Potassium hydroxide. Dissolve 160 g of potassium hydroxide in 100 ml of water.

Ferric chloride, 0.2% w/v in ethanol. Prepare and keep in the dark.

α,α'-Bipyridyl, 0.5% w/v in ethanol.

Ethanol. Redistil highest purity ethanol from 2 g of potassium permanganate and 4 g of potassium hydroxide per litre.

Diethyl ether, peroxide free.

Quinol.

Benzene. This is toxic, handle with care.

Sodium fluorescein, 0.2% w/v in ethanol.

Silica gel G. Thin-layer chromatographic grade 10–40 μ (that supplied by E. Merck A.G., Darmstadt, is suitable).

Petroleum ether, boiling point range 40–60°C.

Standard for T.L.C. α-Tocopherol, 1 μg/μl in ethanol, β-tocopherol, 1μg/μl in ethanol. Dissolve 10 mg of α-tocopherol and 10 mg of β-tocopherol, each in 10 ml of ethanol.

Apparatus

T.L.C. plates prepared as follows:

1. Add 120 ml water and 0.8 ml sodium fluorescein solution to 60 g silica gel.

2. Mix by shaking for 8 minutes.

3. Pour mixture into a variable depth spreader set at 0.75 mm.

4. Spread mixture over glass plates.

5. Allow plates to set for 10 minutes.

6. Dry in an air oven at 100°C for 45 minutes.

7. Store the plates at 30°C until required.

Micrometer syringe.

Ultraviolet lamp.

Spectrophotometer, visible range.

Glass cuvettes, optical pathway 1 cm.

Automatic pipette, 2 ml capacity.

Procedure

(*a*) *Saponification and extraction*:

1. Weigh an amount of sample (but not more than 25 g) containing not less

than 50 μg of α-tocopherol into a 250 ml flat-bottomed flask (Note 1).
2. Continue the saponification and extraction as in 'The determination of vitamin A' instruction (a)1 up to and including (a)14 (Method 7.1).
3. Transfer the residue with two 5 ml washings to a 20 ml tube.
4. Evaporate to dryness on a steam bath under an inert gas.
5. Cool the tube immediately by immersion in cold water.
6. Add by automatic pipette 2 ml of benzene.

(b) *Thin layer chromatographic separation (in subdued light):*
1. In a well ventilated area, apply 4×40 μl spots to a T.L.C. plate, using a micrometer syringe.
2. Apply 10 μl (10 μg) of α-tocopherol and 10 μl (10 μg) of β-tocopherol to the plate to act as a marker. Allow to dry.
3. Develop the plate in a tank containing 15% diethyl ether in petroleum ether for 45 minutes.
4. Allow the plate to dry at room temperature.
5. View the chromatogram under ultraviolet light and mark the position of the tocopherol bands that correspond to the α-band and the β-band.
6. Scrape each band into a 10 ml tube containing 5 ml of ethanol.
7. Shake the tube well and leave to stand for 10 minutes.
8. Scrape and extract a similar area of plate to the tocopherol band area, for a plate blank.
9. Centrifuge the tube for 1 minute.
10. Pipette 3 ml aliquots of the supernatant into 10 ml tubes.

(c) *Spectrophotometric determination (to be done in a dark room):*
1. Set the spectrophotometer to zero at 520 nm using ethanol as reference solution.
2. Add 0.5 ml of α,α'-bipyridyl solution to each tube.
3. Add 0.5 ml of ferric chloride solution to each tube, mix and transfer to a 1 cm glass cuvette.
4. Exactly 2 minutes after the ferric chloride addition, measure the absorbance.

Calculation

Let: Weight (g) of the sample $= W$
            Absorbance of test solution $= a$
            Absorbance of plate blank $= b$
Spectrophotometric factor for 4 ml solution $= F$ (Note 2)
Biological activity factor $= G$ (Note 3)

Calculate α- and β-tocopherol individually using the following calculation:

$$\frac{(a-b) \times F \times G \times 10\ 000}{W \times 3 \times 160} = 20.8 \times \frac{(a-b) \times F \times G}{W}$$

Then: Vitamin E content $(\mu g/g)$ =
Sum of α-tocopherol and β-tocopherol content $(\mu g/g)$.

Notes
1. For samples such as milk the fat is first extracted by the Roese–Gottlieb method. The extracted fat is then saponified.

2. The spectrophotometric factors F are: 98 for α-tocopherol; 96 for β-tocopherol. The spectrophotometric factor multiplied by the absorbance gives micrograms of tocopherol.
3. The biological factors G are: 1 for α-tocopherol; 0.3 for β-tocopherol.

Water-Soluble Vitamins

8.1 Vitamin B$_1$ (Manual Method)

Application

The method is generally applicable to foodstuffs.

Principle

Thiamin is extracted by treatment with hot dilute acid followed by digestion with phosphatase at pH 4.5. The extract is chromatographed on base exchange silicate and the eluted thiamin is oxidised by alkaline ferricyanide to thiochrome, which is then measured fluorimetrically.

Reagents

Hydrochloric acid, 0.2N solution. Dilute 18 ml of hydrochloric acid (sp. gr. 1.18) with water to 1 litre.

Sulphuric acid, 0.1N solution. Dilute 2.8 ml of sulphuric acid (sp. gr. 1.84) to 1 litre with water.

Acetic acid, 3% v/v. Dilute 30 ml of glacial acetic acid (sp. gr. 1.05) to 1 litre with water.

Potassium chloride solution, 25% (w/v) solution in water.

Acid potassium chloride solution. Dilute 8.5 ml of hydrochloric acid (sp. gr. 1.18) to 1 litre with 25% potassium chloride solution.

Sodium hydroxide, 15% w/v aqueous solution.

Alkaline potassium ferricyanide. Dilute 3 ml of 1% potassium ferricyanide solution to 100 ml with 15% sodium hydroxide. Prepare immediately before use.

Sodium acetate, 2.5M solution. Dissolve 205 g (anhydrous) or 340 g (hydrated) of sodium acetate and dilute to 1 litre.

Ethanol. Highest purity, giving no fluorescence at 435 nm on the addition of an equal volume of 15% sodium hydroxide. Redistil ethanol from alkaline potassium permanganate if necessary.

Isobutanol (water saturated). Redistil isobutanol in all-glass apparatus. Add distilled water and shake until saturated.

Enzyme suspension. Suspend with thorough shaking 6 g of a suitable phosphatase source in 2.5M sodium acetate solution and dilute to 100 ml with sodium acetate (Note 1).

Activated base-exchange silicate. A granular artificial zeolite of 60–90 mesh (Note 2). Activate the base exchange silicate as follows:
1. Weigh 500 g of silicate into a 1 litre beaker
2. Add sufficient dilute acetic acid to cover the silicate.
3. Heat at 100°C for 15 minutes, with frequent stirring.
4. Decant the supernatant liquid.
5. Wash the residue by decantation with three portions of hot acid potassium chloride solution.
6. Wash the silicate with boiling water until the washings are chloride-free.

Thiamin stock standard solution, 100 μg/ml. Dissolve 50 mg of thiamin hydrochloride in 0.2N hydrochloric acid and dilute to 500 ml with 0.2N hydrochloric. Store below 5°C; under these conditions the solution keeps for several months.

Thiamin working standard solution, 0.05 μg/ml. Dilute 5 ml of concentrated standard solution, after allowing to warm to room temperature, to 100 ml with 0.2N hydrochloric acid. Pipette 5 ml of this solution into a volumetric flask containing 400 ml of acid potassium chloride and dilute to 500 ml with distilled water. Store below 5°C. Under these conditions the solution keeps for 1 week.

Bromocresol green solution, 0.1% in 20% alcohol.

Apparatus

Chromatographic column. 10 mm internal diameter, length 20 cm, with standard ground socket at top and sintered disc support, fitted with stopcock. A reservoir to contain at least 30 ml may be attached to the upper part of the tube (Fig. 19).

Oxidation vessels. Stoppered separating funnels of 100 ml capacity.

Spectrofluorimeter. Excitation wavelength 360 nm, emission wavelength 435 nm. A Perkin-Elmer 203 with xenon lamp is suitable. If a mercury lamp source is used the excitation wavelength should be 365 nm.

Water bath. Boiling.

Procedure

(*a*) *Extraction*:
1. Weigh a suitable amount of sample (not more than 5 g, and estimated to contain not more than 50 μg of thiamin) into a 100 ml test-tube with a glass stirring rod.
2. Add 50 ml of 0.2N hydrochloric acid.

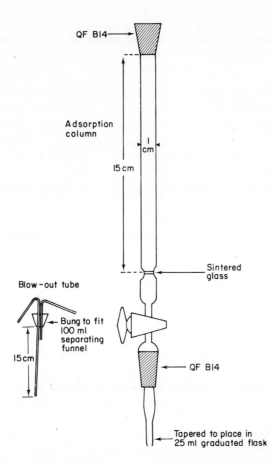

FIG. 19. Chromatographic column for thiamin analysis.

3. Heat the tube for 30 minutes in boiling water, stirring frequently.
4. Check that the sample at the end of digestion is acid to bromocresol green indicator. If not, discard the extract and repeat the digestion with more concentrated acid.
5. Cool the extract to below 50°C.
6. Adjust the pH to between 4 and 4.5 by the addition of sodium acetate solution (5 ml is usually sufficient).
7. Add 5 ml of freshly prepared enzyme solution and mix.
8. Incubate at 37°C overnight.
9. Cool the extracts to room temperature.
10. Transfer the extract to a 100 ml graduated flask, and dilute to volume with water.

11. Filter the extract through a fast filter paper. If necessary dilute this extract to give a solution containing not more than 0.05 μg of thiamin/ml.

(b) *Purification*:

1. Fill the chromatographic column with 6 g of base exchange silicate suspended in water.
2. Allow the water to drain to within 2 mm of the surface of the silicate.
3. Add 5 ml of 3% acetic acid to the column and allow to drain as before.
4. Pipette 25 ml of sample extract on to the column and allow to drain.
5. Discard the eluate.
6. Wash the column with three successive 10 ml portions of boiling water, and discard all the eluates.
7. Elute the thiamin from the column, with two successive 10 ml portions of almost boiling potassium chloride solution from a supply kept hot in a boiling water bath.
8. Collect the eluate in a 25 ml graduated cylinder.
9. Cool the eluate to room temperature, dilute to 25 ml with acid potassium chloride solution, and mix well.

(c) *Oxidation of thiamin*:

1. Pipette two 10 ml aliquots of sample eluate into two 100 ml separating funnels.
2. To one funnel add 5 ml of alkaline potassium ferricyanide (test solution), and to the other 5 ml of 15% sodium hydroxide (blank solution).
3. To both add 5 ml of ethanol, mix well.
4. Add 25 ml of water-saturated isobutanol.
5. Mix by bubbling nitrogen vigorously through solutions for 2 minutes.
6. Remove the lower aqueous layer from each solution leaving 1 ml in the funnel, using the blow-out tube (Fig. 19).
7. Add 2 ml of ethanol to the isobutanol layer, stir carefully, ensuring that the aqueous layer is not disturbed (Notes 3 and 4).
8. Pipette two 10 ml aliquots of thiamin working standard solution into two 100 ml separating funnels. Repeat instructions c.2–c.7.

(d) *Measurement of fluorescence*:

1. Set the excitation at 360 nm and the emission at 435 nm.
2. Fill one fluorimeter cell with 0.1N sulphuric acid and another with an aliquot of the isobutanol layer from the standard determination (c.8).
3. Adjust the fluorimeter to give zero deflection gainst 0.1N sulphuric acid and a deflection of 100 against the standard.
4. Measure the remaining samples and blanks.

Calculation

Let:	Weight (g) of sample	$= W$
	Volume (ml) of extract	$= 100$
	Aliquot (ml) taken for oxidation	$= 10$
	Sample reading	$= T_s$
	Sample blank reading	$= T_b$
	Standard reading	$= S_s$
	Standard blank reading	$= S_b$
	Fluorescence due to thiochrome in sample	$= T_s - T_b$
	Fluorescence due to thiochrome in standard	$= S_s - S_b$

Then: Thiamine content (mg/100 g) of sample
$$= \frac{T_s - T_b}{S_s - S_b} \times \frac{0.5}{W}$$

Notes

1. Taka-diastase supplied by Serva (Heidleberg, Germany) is suitable.
2. Zerolit S/F (Decalso F) ion-exchange resin Permutit Company Ltd. supplied by BDH (Poole) is suitable.
3. Measurement of the fluorescence of the isobutanol should be made within 10 minutes.
4. Avoid exposure of the isobutanol solution to bright daylight as this causes destruction of the thiochrome.

Reference
  Analyst (1951). **76**, 127.

8.2 Vitamin B₂ (Manual Method)

Application
  The method is applicable to all foods.

Principle
  Riboflavin is extracted by treatment with hot dilute acid. After adjusting to pH 4.5, the extract is filtered and decolorised with potassium permanganate and hydrogen peroxide. The fluorescence of the decolorised solution is measured before and after reduction with dithionite solution, the difference in fluorescence being proportional to riboflavin content.

Reagents
  Hydrochloric acid, 0.2N. Dilute 18 ml of hydrochloric acid (sp. gr. 1.18) with water to 1 litre.
  Sulphuric acid, 0.1N. Dilute 28 ml of sulphuric acid (sp. gr. 1.84) with water to 1 litre.
  Hydrochloric acid, 1N. Dilute 80 ml of hydrochloric acid (sp. gr. 1.18) with water to 1 litre.

Sodium hydroxide, 40%. Dissolve 40 g of sodium hydroxide in water and dilute to 100 ml.

Glacial acetic acid, sp. gr. 1.05.

Potassium permanganate, 3%. Dissolve 3 g of potassium permanganate in distilled water and dilute to 100 ml.

Hydrogen peroxide, 3%.

Sodium dithionite $(Na_2S_2O_6.2H_2O)$, solid.

Riboflavin stock standard solution, 25 μg/ml. Dissolve 25 mg of riboflavin in 3 ml of glacial acetic acid, warm if necessary, and dilute to 1 litre with water, avoiding exposure to strong light.

Riboflavin working standard solution, 0.1 μg/ml. Dilute 4 ml of riboflavin strong standard solution to 1 litre with water avoiding exposure to strong light.

Apparatus

pH meter.

Spectrofluorimeter. Excitation wavelength 440 nm, emission wavelength 525 nm. A Perkin-Elmer 203 with xenon lamp is suitable. If a mercury lamp source is used the excitation wavelength should be 436 nm.

Autoclave. A suitable apparatus to provide steam at a pressure of 15 lb/in^2 or 2 atm (i.e. at 120°C).

Procedure

(*a*) *Extraction*:
1. Weigh a suitable amount of sample (not more than 5 g, and estimated to contain not less than 10 μg of riboflavin) into a 100 ml conical flask.
2. Add 50 ml of 0.2N hydrochloric acid and cover the flask with an inverted beaker.
3. Autoclave the sample at 15 lb/in^2 (i.e. 121°C) for 15 minutes.
4. Cool the sample and wash into a 100 ml beaker.
5. Add 40% sodium hydroxide dropwise, while stirring, to the sample extract until it has pH 6.8.
6. Add 1N hydrochloric acid dropwise, whilst stirring, to the sample extract until it has pH 4.5.
7. Transfer the extract to a 100 ml graduated flask and dilute to volume with water.
8. Filter the extract through a fast filter paper. If necessary, dilute this extract to give a solution containing approximately 0.1 μg of riboflavin/ml.

(*b*) *Decolorisation of the extract:*
1. Pipette four 10 ml aliquots of the sample solution into four 25 ml test tubes.

2. Add 2 ml of dilute standard riboflavin solution to two tubes.
3. Add 2 ml of water to the other two tubes.
4. Add 1 ml of glacial acetic acid to each tube and mix.
5. Add 0.5 ml of 3% potassium permanganate solution and mix.
6. Two minutes later, add 0.5 ml of 3% hydrogen peroxide solution and mix.
7. Measure the fluorescence as soon as possible.

(c) *Measurement of fluorescence:*
1. Adjust the fluorimeter to give a deflection of 0 against 0.1N sulphuric acid and 100 against the sample + standard solution at wavelength 525 nm.
2. Measure the fluorescence from each tube in turn.
3. Then add 20 mg of sodium dithionite to each tube in turn and mix. Read the blank value within 10 seconds.

Calculation

Let: Weight (g) of sample $= W$
       Volume (ml) of extract $= 100$
       Aliquot (ml) taken for decolorising $= 10$
       Reading of sample + standard $= T$
       Reading of sample $= S$
       Reading of sample blank $= S_b$
       Reading of sample + standard blank $= T_b$
Hence: Sample + standard fluorescence $= T - T_b$
       Sample fluorescence $= S - S_b$
       Standard fluorescence (0.2 μg) $= (T - T_b) - (S - S_b)$
Then: Riboflavin content (mg/100 g) of sample

$$= \frac{(S - S_b)}{(T - T_b) - (S - S_b)} \times \frac{0.2}{W} \text{ (Note 2)}$$

Notes
1. The whole procedure should take place as quickly as possible in subdued light.
2. If the extract [step (a)8] is diluted (dilution factor $= F$) multiply the result by a factor F.

Reference
Official Methods of Analysis of the Association of Official Agricultural Chemists, 10th Ed., 762 (1965).

8.3 Vitamin B₁ (Thiamin) (High Performance Liquid Chromatographic Method)

Application
The method can be applied to all food products.

Principle
Vitamin B_1 is extracted from foods by acid hydrolysis. Treatment with papain removes interference from proteins, and diastase liberates the vitamin bound in the phosphate form. The clear filtrate is then applied either manually or by means of a fully automated system to a high performance liquid chromatography (HPLC) column. Vitamin B_1 after elution is oxidised by potassium ferricyanide to thiochrome and determined fluorimetrically.

Reagents
Sulphuric acid, 0.25N. Dilute 7 ml of sulphuric acid (sp. gr. 1.84) to 1 litre with water.

Buffer solution. Dissolve 160 g of sodium hydroxide in water and dilute to 500 ml. Dissolve 272 g of glacial acetic acid (sp. gr. 1.04) in water and dilute to 500 ml. Mix equal volumes of these solutions.

Taka diastase. Supplier Serva, Heidelberg. Prepare a suspension containing 100 mg of diastase per ml (Note 1).

Papain. 12 000 E/g, supplier Merck, Darmstadt. Prepare a suspension containing 100 mg of papain per ml (Note 1).

Trichloroacetic acid. Dissolve 45 g of trichloroacetic acid in water and dilute to 100 ml.

Merckosorb SI 60. Fine silica gel for HPLC, particle size 10 micron. Supplier Merck, Darmstadt.

Eluting solution, pH 6.5. Dissolve 11.88 g of disodium hydrogen phosphate dihydrate ($Na_2HPO_4.2H_2O$) in water and dilute to 1 litre. Dissolve 9.08 g of potassium dihydrogen phosphate (KH_2PO_4) in water and dilute to 1 litre. Mix 150 and 350 ml of these solutions, dilute to 1 litre with water, and add 120 ml ethanol.

Akaline ferricyanide solution. Prepare fresh every day. Dissolve 0.5 g of potassium ferricyanide in 50 ml of water. Dissolve 15 g of sodium hydroxide in water, cool to room temperature, and adjust to 100 ml. Combine 24 ml of sodium hydroxide solution, 1 ml of potassium ferricyanide solution, and 25 ml of water.

Thiamin stock standard solution, 500 μg/ml. Dissolve 125 mg of thiamin hydrochloride in water, add 20 ml of 0.25N sulphuric acid and 2 ml of 45% trichloroacetic acid solution. Dilute to 250 ml in a graduated flask with water. Store in the dark at 0–4°C.

Thiamin working standard solution, 4 μg/ml. Pipette 2 ml of thiamin stock standard solution into a 250 ml graduated flask. Dilute to volume with water.

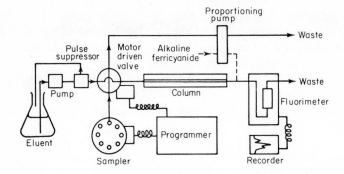

FIG. 20. Scheme of automated liquid chromatography for thiamin assay.

Apparatus

Autoclave. A suitable apparatus to provide steam at a pressure of 15 lbs/sq. in. (or 2 atm. i.e. at 121°C).

Waterbath. Preferably provided with a shaking device, at 40–45°C and 50–60°C.

High speed centrifuge, 30 000*g* maximum.

Centrifuge tubes, 50 ml capacity.

Apparatus for high performance liquid chromatography. A schematic diagram of a fully automated apparatus is shown in Fig. 20, the main parts of which are:

1. High pressure pump.
2. Sampler with sample vessels, volume 1.7 ml.
3. Steel column, 12.5 cm × 0.46 cm i.d.; column packing Merckosorb SI 60, particle size 10 micron.
4. Sample loop, 60 μl.
5. Fluorimetric detection system and flow through cell; excitation 366 nm, emission 464 nm.
6. Proportioning pump.

Procedure (Note 2)

1. Weigh out accurately, into a 50 ml centrifuge tube, sufficient sample to give 0.5 g of solid material (e.g. a sample containing about 80% water, weigh 2.50 g).
2. Add water by graduated pipette to the tube such that the total water content is 2 g (e.g. a sample containing less than 10% water would require the addition of 2 ml).
3. Pipette 2 ml of thiamin working standard solution into a centrifuge tube (N.B. for every five sample tubes have one standard tube).
4. Add to each tube 10 ml of 0.25N sulphuric acid.

5. Cover the tube with aluminium foil.
6. Heat in an autoclave at 121°C (15 lb/in²) for 30 minutes.
7. Cool the tube, add 1.5 ml of buffer solution and mix to give pH 4.6.
8. Add 1 ml of takadiastase suspension and mix well.
9. Place the tube in a water bath at 40–45°C for 25 minutes. Shake the tubes during this period.
10. Add 1 ml of papain solution, mix well, and continue heating for 2 hours at 40–45°C.
11. Add 2 ml of 45% trichloroacetic acid, mix well.
12. Heat the tube for 5 minutes in a water bath at 50–60°C.
13. Centrifuge the tube for 5 minutes at $30\,000g$.
14. Either inject 60 μl of the supernatant manually on to the column or fill the sample vessels of the fully automated apparatus.
15. Chromatograph at a flow rate of 1 ml per minute and add alkaline ferricyanide reagent at a flow rate of 0.3 ml per minute (see Fig. 20).
16. Measure the peak heights of samples and standards on the chromatograms obtained.

Calculation

Let: Height of the sample peak $= h$
      Height of the standard peak $= H$
      Weight (g) of sample $= W$
Then: Thiamin (mg/100 g) in sample (expressed as thiamin chloride HCl)

$$= (8 \times h \times 100)/(W \times H \times 1000)$$
$$= 0.8h/W \times H.$$

Notes
1. Check each batch of enzymes through the method by using cocarboxylase as substrate. If these reagents have a blank value correct the height of the sample and standard peaks.
2. During the whole procedure the samples should be protected against direct ultraviolet light.

8.4 Vitamin B₂ (High Performance Liquid Chromatographic Method)

Application
    The method can be applied to all food products.

Principle
    Vitamin B_2 is extracted from foods by acid hydrolysis. Treatment with papain removes interference from proteins, and diastase liberates the vitamin bound in the phosphate form. The clear filtrate is then applied either manually

or by means of a fully automated system to a high performance liquid chromatography (HPLC) column. Vitamin B_2 is determined directly by fluorimetry.

Reagents

Sulphuric acid, 0.25N. Dilute 7 ml of sulphuric acid (sp. gr. 1.84) to 1 litre with water.

Buffer solution. Dissolve 160 g of sodium hydroxide in water and dilute to 500 ml. Dissolve 272 g of glacial acetic acid (sp. gr. 1.04) in water and dilute to 500 ml. Mix equal volumes of these solutions.

Taka diastase. Supplier Serva, Heidelberg. Prepare a suspension containing 100 mg of diastase per ml (Note 1).

Papain. 12 000 E/g, supplier Merck, Darmstadt. Prepare a suspension containing 100 mg of papain per ml (Note 1).

Trichloroacetic acid. Dissolve 45 g of trichloroacetic acid in water and dilute to 100 ml.

Merckosorb SI 60. Fine silica gel for HPLC, particle size 10 micron supplied by Merck, Darmstadt.

Eluting solution, pH 4.6. Dissolve 27.2 g of sodium acetate trihydrate $(CH_3COONa.3H_2O)$ in water and dilute to 1 litre. Dilute 12 g of acetic acid to 1 litre. Mix equal volumes of these solutions and dilute 5 times.

Riboflavin stock standard solution, 30 μg/ml. Weigh 15.0 mg of riboflavin and transfer with 300 ml of warm water (50°C) to a 500 ml graduated flask, add 40 ml of 0.25N sulphuric acid solution and 4 ml of 45% trichloroacetic acid solution. Swirl until the riboflavin is dissolved and dilute to 500 ml with water. Store in the dark at 0–4°C.

Riboflavin working standard solution, 1.5 μg/ml. Pipette 10 ml of riboflavin stock standard solution into a 200 ml graduated flask. Dilute to volume with water.

Apparatus

Autoclave. A suitable apparatus to provide steam at a pressure of 15 lb/in^2 or 2 atm (i.e. at 120°C).

Water bath, preferably provided with a shaking device, at 40–45°C and 50–60°C.

High speed centrifuge, 30 000g maximum.

Centrifuge tubes, 50 ml capacity.

Apparatus for high performance liquid chromatography. A schematic diagram of a fully automated apparatus is shown in Fig. 21 the main parts of which are:

1. High pressure pump.
2. Sampler with sample vessels, volume 1.7 ml.

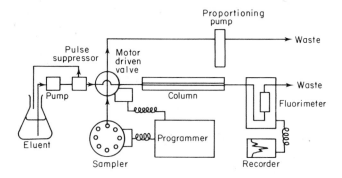

FIG. 21. Scheme of automated liquid chromatography for riboflavin assay.

3. Steel column, 25 cm × 0.46 cm i.d.; column packing Merckosorb SI 60, particle size 10 μm.
4. Sample loop, 60 μl.
5. Fluorimetric detection system and flow through cell, excitation 435 nm, emission 545 nm.
6. Proportioning pump.

Procedure (Note 2)
1. Weigh out accurately, into a 50 ml centrifuge tube, sufficient sample to give 0.5 g of solid material (e.g. a sample containing about 80% water, weigh 2.5 g).
2. Add water by graduated pipette of the tube such that the total water content is 2 g (e.g. a sample containing less than 10% water would require the addition of 2 ml).
3. Pipette 2 ml of riboflavin working standard solution into a centrifuge tube (N.B. for every five sample tubes include one standard tube).
4. Add to each tube 10 ml of 0.25N sulphuric acid.
5. Cover the tube with aluminium foil.
6. Heat in an autoclave at 121°C (15 lb/in²) for 30 minutes.
7. Cool the tube, add 1.5 ml of buffer solution, and mix to give pH 4.6.
8. Add 1 ml of takadiastase suspension and mix well.
9. Place the tube in a water bath at 40–45°C for 25 minutes. Shake the tubes during this period.
10. Add 1 ml of papain solution, mix well, and continue heating at 40–45°C for a further 2 hours.
11. Add 2 ml of 45% trichloroacetic acid, mix well.
12. Heat the tube for 5 minutes in a water bath at 50–60°C.
13. Centrifuge the tube for 5 minutes at 30 000g.

14. Either inject 60 μl of the supernatant manually on to the column or fill the sample vessels of the fully automated apparatus.
15. Chromatograph at a flow rate of 1 ml per minute.
16. Measure the peak heights of samples and standards.

Calculation

Let: Height of the sample peak $= h$
     Height of the standard peak $= H$
     Weight (g) of sample $= W$
Then: Riboflavin (mg/100 g) $= (3/W) \times (h/H) \times (100/1000)$
                            $= 0.3h/W \times H.$

Notes
1. Check each batch of enzymes through the method by the hydrolysis of riboflavin mono-5-phosphate-Na. If these reagents have a blank value correct the height of the sample and standard peaks.
2. During the whole procedure the samples should be protected against ultraviolet light. Carry out the procedure in a darkened room.

8.5 Niacin (Automated High Performance Liquid Chromatographic Method)

Application
The method can be applied to all food products.

Principle
Niacin is extracted from foods by acid hydrolysis. Treatment with papain releases vitamin from proteins, and diastase liberates the vitamin bound in the phosphate form. The clear filtrate is then applied either manually or by means of a fully automated system to high pressure liquid chromatography column. Nicotinic acid and nicotinamide are separated and individually determined after reaction, with cyanogen bromide and *p*-aminoacetophenone by fluorimetry.

Reagents
Sulphuric acid, 0.25N. Dilute 7 ml of sulphuric acid (sp. gr. 1.84) to 1 litre with water.
Buffer solution. Dissolve 160 g of sodium hydroxide in water and dilute to 500 ml. Dissolve 272 g of glacial acetic acid (sp. gr. 1.04) in water and dilute to 500 ml. Mix equal volumes of these solutions.
Taka diastase. Supplier Serve, Heidelberg. Prepare a suspension containing 100 mg of diastase per ml (Note 1).
Papain, 12000 E/g. Supplier, Merck, Darmstadt. Prepare a suspension containing 100 mg of papain per ml (Note 1).

Trichloracetic acid, 45%. Dissolve 45 g of trichloroacetic acid in water and dilute to 100 ml.

Merkosorb SI 60. Fine silica gel for HPLC particle size 10 microns. Supplied by Merck, Darmstadt.

Eluting solution. Dissolve 27.2 g of sodium acetate trihydrate in water and dilute to 1 litre. Dissolve 12 g of acetic acid in water and dilute to 1 litre. Mix equal volumes of these two solutions and dilute five times.

Cyanogen bromide solution. Dissolve 5.0 g of cyanogen bromide in 100 ml of water. This should be done in a fume cupboard (Note 2).

p-Aminoacetophenone. Dissolve 5.0 g of p-aminoacetophenone in 100 ml of 2N hydrochloric acid.

Niacin stock standard solution, 2400 μg/ml. Weigh 120 mg of nicotinic acid and 120 mg of nicotinamide and transfer with 60 ml of water into a 100 ml graduated flask. Add 10 ml of 0.25N sulphuric acid and 2 ml of 45% trichloroacetic acid solution then dilute to volume with water.

Niacin stock standard solution, 60 μg/ml. Pipette 5 ml of niacin stock standard solution into a 200 ml graduated flask and dilute to volume with water.

Apparatus

Autoclave. A suitable apparatus to provide steam at a pressure of 15 lb/in^2 (i.e. at 121°C).

Water baths. Preferably provided with shaking device. Temperature 40–45°C and 50–60°C.

High speed centrifuge, 30 000g maximum.

Centrifuge tubes, 50 ml capacity.

Apparatus for high performance chromatography. A schematic diagram of a fully automated apparatus is shown in Fig. 22. Comprising:
1. High pressure pump.
2. Sampler with sample vessels, volume 1.7 ml.
3. Steel column, 25 cm × 0.46 cm i.d.
4. Sample loop, 60 μl.
5. Fluorimetric detector and flow through cell, excitation 435 nm, emission > 500 nm.
6. Analyser system (Fig. 23).

Procedure
1. Weigh out accurately, into a 50 ml centrifuge tube, sufficient sample to give 0.5 g of solid material (e.g. a sample containing about 80% water, weigh 2.50 g).
2. Add water by graduated pipette to the tube to give total water content of 2

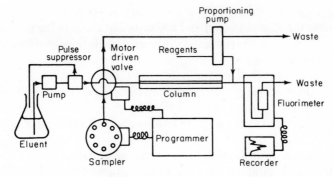

FIG. 22. Scheme of automated liquid chromatography for niacin assay.

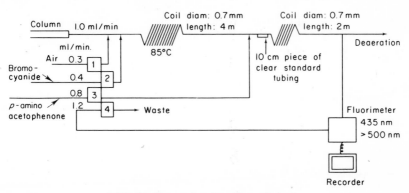

FIG. 23. Set-up for detection of niacin.

ml (e.g. a sample containing 10% water would require the addition of 2 ml).

3. Pipette 2 ml of niacin working standard solution into a centrifuge tube (N.B. for every five sample tubes include one standard tube: the standard solutions are not given the heat treatments as described in steps 6, 9, 10, and 12).

4. Add to each tube 10 ml of 0.25N sulphuric acid.

5. Cover the tubes with aluminium foil.

6. Heat in an autoclave at 121°C and 15 lb/in² for 30 minutes.

7. Cool the tubes, add 1.5 ml of buffer solution and mix to give pH 4.6.

8. Add 1 ml of Taka diastase suspension and mix well (Note 3).

9. Place the tubes in a water bath at 40–45°C for 25 minutes. Shake the tubes during this period.

10. Add 1 ml of papain solution, mix well, and continue heating tubes at 40–45°C for a further 2 hours. (Note 3).

11. Add 2 ml of 45% trichloracetic acid, mix well.

12. Heat tubes for 5 minutes in a water bath at 50–60°C.
13. Remove the aluminium foil and add sufficient water to the sample tube to make the weight of its contents 0.5 g more than the weight of the contents of the standard tubes.
14. Centrifuge the tubes for 5 minutes at 30 000g.
15. Either inject 100 μl of the supernatant onto the column or fill the sample vessels of the fully automated apparatus.
16. Chromatograph at a flow rate of 1 ml/minute and add reagents as shown in Fig. 23.
17. Measure the peak heights for nicotinic acid and nicotinamide of both samples and standards on the chromatograms obtained.

Calculation

Let: Weight (g) of sample taken $= W$
      Weight of nicotinic acid peak in sample $= a$
      Weight of nicotinamide in sample $= b$
      Weight of nicotinic acid in standard $= A$
      Weight of nicotinamide in standard $= B$
Then: Nicotinic acid (mg/100 g) $= [a/(W \times A)] \times 6$
      Nicotinamide (mg/100 g) $= [b/(W \times B)] \times 6$
and   Total Niacin (mg/100 g) $= [(a/A)+(b/B)] \times (6/W)$

Note (See Note added in proof, p. 238.)

8.6 Introduction to Microbiological Assay Techniques for the Determination of Vitamins

Principle
Microbiological methods are based on the fact that certain micro-organisms require specific vitamins for growth. If a basal medium complete in all respects except for the vitamin under test is used, the growth response of the organism in standard and unknown solutions can be compared quantitatively.

Analytical area
The area for microbiological techniques should be a self-contained area, the windows fitted with blinds to eliminate daylight. It should be of structural simplicity to facilitate cleaning. After each assay the benches should be wiped with a dilute solution of Chloros.

Apparatus
Autoclave, capable of accurate adjustment to a constant pressure of 15 lb/in^2 (121°C).

Incubator, which will maintain constant and uniform temperature ($\pm 0.5^{\circ}$C) in the range 28–37°C.

Refrigerator.

pH-Meter.

Nephelometer.

Vortex mixer.

Pipette washer. A continuous syphoning device with a removable inner basket.

Culture tube racks.

Automatic dispenser, 5 and 10 ml capacity. Zippettes are suitable.

Inoculating needle and loop.

Assay tubes, lipless, size as specified in methods, fitted with polypropylene caps (different colour for each method).

Pipettes, sizes as specified in methods.

Cleaning reagents

Hypochlorite solution. Chloros; a commercial solution is suitable.

Detergent. Pyroneg; a specialised product for cleaning glassware is suitable.

Glass Cleaning

Pipettes. Soak for 16 hours in a solution of Chloros at a concentration of 300 p.p.m. of free chlorine. Soak for a further 16 hours in 50% v/v hydrochloric acid. Transfer to pipette washer and rinse thoroughly.

Glassware. Soak dirty glassware in a water bath (at 60°C) containing Pyroneg detergent (100 g dissolved in 20 litres water) for 16 hours. Rinse each piece of glassware three times with hot water, three times with cold water, and finally once in distilled water. Dry in an oven.

Media

Commercially supplied media (e.g., from Difco) are suitable for stock cultures and for the assays (details are given in Maintenance of Stock Cultures and in each method).

Maintenance of stock cultures

Test organisms

Niacin *Lactobacillus plantarum*
        in freeze dried form from the N.C. IB no. 6376

B6   *Saccharomyces carlsbergensis*
        in freeze dried form from N.C.Y.C. no. 74

B12   *Lactobacillus leichmanii*
        in freeze dried form from the N.C.1B no 8118

B2   *Lactobacillus casei* ATCC no. 7469

Lactobacillus plantarum, leichmanii, and *casei* are maintained on 10 ml slopes of M.R.S. agar in a 1 oz McCartney bottle. The bacterium is subcultured weekly by transferring aseptically a loop of bacteria from the current culture onto a fresh slope. This is incubated at 30° overnight and can be stored at 0° in a refrigerator for up to 1 month.

Saccharomyces carlsbergensis is a yeast maintained on 10 ml agar slopes using *Lactobacillus* Agar AOAC. It is subcultured weekly and kept under the same conditions as *Lactobacillus plantarum.*

Water
Use good quality distilled water throughout the assays.

8.7 Vitamin B$_2$ (Riboflavin) (Microbiological Method)

Application
The method is generally applicable to foodstuffs.

Principle
Riboflavin is extracted from the foodstuff using acid to hydrolyse bound forms. After suitable dilution the riboflavin content is estimated nephelometrically by the growth of *Lactobacillus casei,* which has a specific growth requirement for riboflavin under carefully controlled conditions.

Reagents
Micro inoculum broth. Difco (0320–02) prepared according to Difco instructions.
Riboflavin assay medium. Difco (0325–15). Use double strength for 6 hour broth tubes and assay of samples and standards. For this strength dissolve 4.8 g of media per 100 ml of water.
Hydrochloric acid, 0.1N.
Sodium hydroxide solution, 1N and 0.1N.
Riboflavin standard solution, 200 µg/ml. Dissolve 68.33 mg of riboflavin 5-phosphoric acid monosodium salt (Note 1) in 50 ml of 0.1N hydrochloric acid and dilute to 250 ml with distilled water. Store in a refrigerator away from daylight. This is solution A.

Test organism
Lactobacillus casei. ATCC no. 7469.

Apparatus
Assay tubes, 19 mm × 150 mm Pyrex heavy duty.
Filter paper, Whatman No. 42 or 541; Schleicher and Schüll 589–3 or 1505.

Procedure

Two extracts of each sample are prepared and each is assayed in duplicate at three concentrations.

(*a*) *Extraction:*

1. Weigh into a 250 ml beaker between 2.0 and 10.0 g of sample, containing not more than 10 μg of riboflavin.
2. Add 50 ml of 0.1N hydrochloric acid.
3. Cover the beakers with aluminium foil and autoclave at 121°C for 30 minutes. (After this stage all operations must be carried out in subdued light.)
4. Remove the samples from the autoclave and cool to room temperature.
5. Add 1N and finally 0.1N sodium hydroxide to adjust the pH to 6.8 and then make up to 100 ml with water.
6. Filter through a No. 42 filter paper, or, if the samples are non-fatty, a No. 541 filter paper.
7. Dilute the filtrate further, if necessary, so that the total riboflavin concentration of the solution to be used for the assay is approximately 40 ng/ml.
8. Take a reagent blank through the same procedure.

(*b*) *Preparation of inoculum:*

1. Subculture the organism overnight from the most recent culture, into 5 ml micro inoculum broth which has been previously sterilised at 121°C for 5 minutes. Cap and incubate overnight at 37°C.
2. The next morning mix the contents of the tube and transfer 1 drop of the broth into 5 ml of double strength B_2 assay medium fortified with B_2 (see Note 3).
3. Incubate for 6 hours at 37°C.
4. Prepare the inoculum by transferring 1 drop with a sterile pipette to an assay tube containing 5 ml of water plus 5 ml of double strength sterilised assay media.

(*c*) *Preparation of assay solutions:*

1. Prepare working standard solution by dilution of standard solution A as follows. (200 μg/ml) 5 ml→500 ml and 4 ml→200 ml to give a working standard solution of 40 ng/ml.
2. Into assay tubes pipette 0, 1, 2, 3, 4, and 5 ml of standard solution C in triplicate.
3. Prepare sample solutions by diluting sample extracts if necessary to give a concentration of approximately 40 ng/ml (see instruction a.7).
4. Into assay tubes pipette 1, 2, and 3 ml of sample solutions in duplicate. Adjust volume of tubes to 5 ml with water.

5. Prepare blanks by dispensing into 6 assay tubes 5 ml of water.
6. Add 5 ml of double strength medium to every tube containing standard, sample or blank.
7. Cap the tubes and autoclave at 121°C for not more than 5 minutes. Cool to room temperature.

(d) *Inoculation and incubation of tubes:*
1. Loosen all caps but leave tubes covered.
2. Fill a sterile 10 ml pipette with the inoculum.
3. Remove cap from first assay tube, then add one drop only of inoculum and rapidly replace cap.
4. Inoculate the remainder of the tubes in a similar manner except for the blank tubes.
5. Place the assay in the incubator for 22–24 hours at 37°C.

(e) *Measurement of growth by nephelometer:*
1. Mix the liquid in the tubes (mix each set of 6 tubes just before reading) using a vortex mixer and then read the assay on an EEL nephelometer using a neutral filter.
2. Set the uninoculated blank at zero on the scale and the top standard at 90–100 per cent full scale deflection.
3. Read the rest of the standards and samples.

Calculation
1. Construct a standard curve by plotting nephelometer readings against ng riboflavin per tube for standards.
2. From the graph obtain the amount of riboflavin (ng) in each pair of assay tubes x_1, x_2, x_3.
Let the aliquots (ml) taken for assay $= v_1, v_2, v_3$ respectively.
Then concentration of riboflavin in sample extract $(x_1/v_1 + x_2/v_2 + x_3/v_3)/3$ ng/ml $= a$ ng/ml.

Let: Weight (g) of sample $= W$
      Volume (ml) of extract $= V$
Then: Riboflavin content (mg/100 g) $= (a \times V \times 10^2)/(W \times 10^6)$
                                    $= (a \times V)/(W \times 10^4)$

Notes
1. The phosphoric acid derivative of riboflavin (flavin mononucleotide) is preferred to riboflavin because of its greatly increased solubility. 68.33 ng of FMN $=$ 50 mg of riboflavin.
2. If a high result is obtained for the uninoculated blanks contamination has occurred and the assay should be repeated.

3. Prepare riboflavin broth tube by adding 0.2 ml of standard Y to 5 ml aliquots of double strength riboflavin assay medium, autoclave for not more than 5 minutes at 121°C. It is important that both the riboflavin medium and the micro-inoculum broth be at 37°C before transfer of the organisms, and this is done by placing both the medium and the broth in the 37°C incubator overnight.

8.8 Niacin (Microbiological Method)

Application
The method is generally applicable to foodstuffs.

Principle
Niacin is extracted from the foodstuff using acid to hydrolyse bound forms. After suitable dilution the niacin content is estimated nephelometrically by the growth of *Lactobacillus plantarum*, which has a specific growth requirement for niacin, under carefully controlled conditions.

Reagents
Micro-inoculum broth. Difco (0320–02) prepared according to Difco instructions.
Niacin assay medium. Difco (0322–15). Use single strength for six hour broth tubes and assay of samples and standards. For this strength dissolve 3.75 g of media per 100 ml of water.
Sodium hydroxide solution, 4N.
Niacin standard solution. Dissolve 250 mg of nicotinic acid in water and dilute to 500 ml. Store in a refrigerator.
Test organism, *Lactobacillus plantarum* (NCIB No. 6376).

Apparatus
Assay tubes; 19 mm × 150 mm Pyrex heavy duty.
Micro-Pipette; 0.2 ml Emil gold-line are suitable.
Mixer.
Filter paper; Whatman No. 42 or 541, Schleicher and Schüll 589–3 or 1505.

Procedure
(a) *Extraction*:
1. Weigh between 2.0 and 10.0 g of sample into a 250 ml beaker, containing not more than 50 μg of niacin.
2. Add 50 ml of 1N hydrochloric acid and stir.

3. Cover the beaker with an aluminium foil cap.
4. Autoclave for 15 minutes at 121°C (15 lb/in²).
5. Remove from autoclave and allow to cool.
6. Add 10 ml of 4N sodium hydroxide by cylinder.
7. Adjust to pH 4.5, using a pH meter, with additions of 1N and 0.1N sodium hydroxide solution.
8. Rinse electrode with water.
9. Transfer contents of beaker to graduated flask and dilute to 100 ml with water.
10. Filter through Whatman No. 42 paper or No. 541 paper if samples do not contain fat.
11. Store the filtrate in the refrigerator overnight.
12. Dilute the filtrate, if necessary, to give a niacin concentration of approximately 400 ng/ml.

(b) *Preparation of assay solutions*:
1. Prepare standard solutions by serial dilution of the stock standard solution (500 µg/ml) 10 ml→1 litre and 10 ml→100 ml to give a working standard solution of 500 ng/ml.
2. Pipette 0, 0.04, 0.08, 0.12, 0.16, and 0.2 ml aliquots in duplicate into assay tubes.
3. Using a suitable dispenser, add 10 ml of single strength assay media into each assay tube containing standard.
4. Prepare sample solutions by pipetting 0.05, 0.1, and 0.2 ml aliquots of each extract in duplicate into assay tubes.
5. Using a suitable dispenser, add 10 ml of single strength assay medium into each assay tube containing sample solution.
6. Dispense 10 ml of media into a further set of 6 assay tubes to act as blanks.
7. Cap each tubed with a polypropylene cap.
8. Autoclave the assay for 15 minutes at 121°C (15 lb/in²).
9. Remove from autoclave and cool to approximately 30°C.

(c) *Preparation of inoculum*:
1. Transfer cells from the *Lactobacillus plantarum* culture to a tube containing 5 ml of sterile micro inoculum broth (Note 1).
2. Incubate for 18 hours at 37°C.
3. Transfer 1 drop of this sub-culture to a tube containing 5 ml of sterile niacin broth (Note 1).
4. Incubate for 6 hours at 37°C.
5. Prepare the inoculum by transferring a few drops of this sub-culture with a sterile pipette to an assay tube containing 10 ml of single strength assay medium.

(d) *Inoculation of tubes*:
1. Fill a sterile 10 ml pipette with the inoculum.
2. Remove the cap from the first assay tube.
3. Add 1 drop of inoculum and rapidly replace cap.
4. Repeat the process for each tube in the assay except the 5 blank tubes (uninoculated blanks).
5. Place the assay into the incubator at 37°C for 22 hours.

(e) *Measurement of growth*:
1. Mix the contents of each assay tube on a Vortex mixer immediately before reading in the nephelometer.
2. Place an uninoculated blank tube in the instrument and adjust the reading to zero.
3. Place an assay tube containing 0.2 ml of standard in the instrument and set the reading to approximately 95–100% full scale deflection.
4. Read the remaining assay solutions, mixing each before placing in the instrument, and record the galvanometer readings.

Calculation
1. Construct a standard curve by plotting nephelometer readings against ng niacin per tube for standards.
2. From the graph obtain the amount of niacin (ng) in each pair of assay tubes x_1, x_2, x_3.
   Let: Aliquots (ml) taken for the assay = v_1, v_2, v_3 respectively.
   Then: Concentration of niacin in sample extract
   $(x_1/v_1 + x_2/v_2 + x_3/v_3)/3$ ng/ml = a ng/ml

Let: Weight (g) of sample   = W
      Volume (ml) of extract   = V
Then: Niacin content (mg/100 g) = $(a \times V \times 10^2)/(W \times 10^6)$
                                = $(a \times V)/(W \times 10^4)$

Notes
1. Niacin broth tubes are prepared by adding 5 ml of single strength niacin assay medium to 0.2 ml of working standard solution (500 ng/ml) contained in 16 mm × 100 mm tubes. These tubes are capped and autoclaved for 5 minutes at 121°C (15 lb/in^2).
2. The response surve is essentially linear but may deviate slightly from linearity at above 80 ng level.

Reference
Laboratory Practice (1961). **10**, 633.

8.9 Vitamin B$_6$ (Microbiological Method)

Application
   The method is generally applicable to foodstuffs.

Principle
   Vitamin B$_6$ is extracted from the foodstuff with acid to hydrolyse bound forms. After suitable dilution the vitamin B$_6$ content is estimated nephelometrically using the growth of the yeast *Saccharomyces carlsbergensis*, which has a specific growth requirement for vitamin B$_6$ under carefully controlled conditions and gives total B$_6$ activity (i.e. responds equally to pyridoxine, pyridoxal, and pyridoxamine) (Note 1).

Reagents
   Pyridoxine Y Medium. Difco (0951–15). Use double strength for assay of samples and standards. For this strength dissolve 5.3 g per 100 ml water.
   Sulphuric acid, 0.44N. Dilute 12.3 ml of concentrated sulphuric acid (sp. gr. 1.84) to 1 litre with water.
   Sodium hydroxide solution, 4N. Dissolve 160 g of sodium hydroxide in water and dilute to 1 litre.
   Pyridoxine stock standard solution (A), 100 μg/ml. Dissolve 122 mg of pyridoxine hydrochloride in 1 litre of 25% ethanol. Store the standard solution in the refrigerator until required. Stable for at least 1 month.
   Test organism, *Saccharomyces carlsbergensis* (ATCC No. 9080, NCYC 74).

Apparatus
   Assay tubes, 19 mm × 150 mm Pyrex heavy duty.
   Micro-pipettes, 0.2 ml E-mil gold line are suitable.
   Filter paper, Whatman No. 42 or 541/ Schleicher and Schüll No. 589–3 or 1505.

Procedure
(a) *Extraction*:
   1. Weigh between 0.5 g and 15 g into a 400 ml beaker, containing not more than 12.5 μg vitamin B$_6$.
   2. Add 180 ml of 0.44N sulphuric acid.
   3. Cover the beaker with an aluminium foil cap.
   4. Autoclave for 5 hours at 121°C.
   5. Remove from autoclave and cool in water bath.
   6. Add 20 ml of 4N sodium hydroxide, using a measuring cylinder.
   7. Adjust to pH 4.5 on a pH meter, using 1N and 0.1N sodium hydroxide solution.

8. Rinse electrode with water.
9. Transfer contents of beaker to volumetric flask and dilute to 250 ml with water.
10. Filter through Whatman No. 42 paper (or No. 541 paper if samples contain no fat).
11. Store the filtrate in the refrigerator until needed (extracts may be stored for up to 36 hours).
12. Dilute the filtrate if necessary to give pyridoxine concentration of approximately 50 ng/ml.

Reagent blanks are taken through the same procedure in duplicate.

(b) *Preparation of assay solution*:
1. Serially dilute the pyridoxine stock standard solution 5 ml→500 ml and 5 ml→100 ml to give a pyridoxine working solution (50 ng/ml). Pipette 0, 0.04, 0.08, 0.12, 0.16, and 0.2 ml of working standard solution in duplicate into assay tubes.
2. Pipette 0.05, 0.10, and 0.20 ml aliquots of each sample extract in duplicate into assay tubes.
3. Using a suitable dispenser add 5 ml of double strength medium to each assay tube.
4. Dispense 5 ml aliquots of medium into a further set of assay tubes to act as blanks.
5. Cap each tube with polypropylene cap.
6. Autoclave the assay for 5 minutes at 121°C (Note 3).
7. Remove from autoclave and cool to below 30°C in water bath.

(c) *Preparation of the inoculum*:
1. Transfer cells of S. *carlsbergensis* from the agar slope (Note 4) to a tube containing 5 ml of sterile double strength pyridoxine Y medium.
2. Incubate for 20 hours at a temperature between 28 and 30°C with the loosely capped tube inclined at an angle of 45°.
3. Prepare inoculum by diluting the subculture aseptically with sterile double strength medium until a reading of 50 is obtained on the standard perspex scale (Note 6).

(d) *Inoculation and incubation of tubes*:
1. Fill a sterile 10 ml pipette with the inoculum.
2. Remove the cap from the first assay tube.
3. Immediately add 1 drop only of the inoculum to the tube and replace the cap loosely.
4. Repeat the process for each tube in the assay except the uninoculated blank tubes.

5. Place the assay in an incubator with each tube in an inclined position.
6. Incubate at 30°C for 2–22 hours.
7. Remove tubes from the incubator and add 7 ml of water to each.

(e) *Measurement of growth*:
1. Mix the contents of each assay tube on a Vortex mixer immediately before reading the nephelometer.
2. Place an uninoculated blank tube in the instrument and adjust the reading to zero.
3. Place an assay tube containing 0.2 ml of standard in the instrument and set the reading to approximately 95–100% full scale deflection.
4. Read the remaining assay solutions, mixing each before placing in the instrument and record the galvanometer readings.

Calculation
1. Construct a standard curve by plotting nephelometer readings against ng pyridoxine per tube for standards.
2. From the graph obtain the amount of pyridoxine (ng) in each pair of assay tubes, x_1, x_2, x_3.
Let: Aliquots (ml) taken for the assay be v_1, v_2, v_3 respectively.
Then: Concentration of pyridoxine in sample extract =
$(x_1/v_1 + x_2/v_2 + x_3/v_3)/3$ ng/ml = a ng/ml.

Let: Weight (g) of sample   = W
      Volume (ml) of extract   = V
Then: Pyridoxine content (mg/100 g) = $(a \times V \times 10^2)/(W \times 10^6)$
                                     = $(a \times V)/(W \times 10^4)$

Notes
1. Although it is claimed by many workers that *Saccharomyces carlsbergensis* responds equally to all three forms of vitamin B_6, it has been found with this method that its response to pyridoxamine can vary from 30 to 60%. Vitamin B_6 results on foods particularly high in pyridoxamine (e.g. meat) could therefore be low.
2. All operations must be carried out under artificial light.
3. Owing to the heat sensitivity of the assay it is recommended that the larger autoclaves having extended cooling-down periods, the time at 121°C (15 lb/in^2) be reduced – 1 minute at this temperature is sufficient.
4. *Saccharomyces carlsbergensis* is a bottom yeast and is maintained under aerobic conditions on an agar slope using 10 ml of sterilised lactobacilli agar in a 1 oz McCartney bottle. The yeast is subcultured weekly by taking a loopful of yeast from the current culture on to a fresh agar slope. It is

incubated at 30°C overnight and stored at 0°C in a refrigerator for up to 1 month.

5. The nephelometer is set to read 0 on a water blank and 100 on the Perspex standard obtained from E.E.L. This is the 'Standard Perspex Scale'.

Reference
Methods of Biochemical Analyses, Vol. 8, Interscience (1960).

8.10 Vitamin B$_{12}$ (Microbiological Method)

Application
The method is generally applicable to foodstuffs.

Principle
Vitamin B$_{12}$ is released from foodstuffs using an acetate buffered solution at pH 4.5. The liberated unstable cobalamins are converted in the presence of sodium cyanide into stable cyanocobalamins. After suitable dilution the vitamin B$_{12}$ content is estimated nephelometrically by the growth of *Lactobacillus leichmanii* which has a specific growth requirement for vitamin B$_{12}$ under carefully controlled conditions.

Reagents
Sodium acetate buffer. Dissolve 2.9 ml of glacial acetic acid (sp. gr. 1.05) and 6.8 g of sodium acetate trihydrate in water and dilute to 500 ml.

Sodium cyanide solution, 1%. Dissolve 1 g in 100 ml of distilled water (POISON—GREAT CARE MUST BE TAKEN).

Vitamin B$_{12}$ assay media. Difco (0457–15). Use single strength for six hour broth tubes and standards. For this strength dissolve 4.25 g of media per 100 ml of water.

Micro inoculum broth. Difco (0320–02) prepared according to instruction on the bottle.

Vitamin B$_{12}$ standard stock solution. Dissolve 10 mg of cyanocobalamin in 1 litre of 25% ethanol–water mixture. Store in a dark bottle and keep in a refrigerator (10 μg/ml). Stable for at least 1 month.

Test organism, *Lactobacillus leichmanii* (NCIB 8118 or ATOC 7830).

Apparatus
Assay tubes, 25 mm × 150 mm and 19 mm × 150 mm Pyrex heavy duty.

Water bath, at 100°C.

Filter paper, Whatman No. 42 or 541, Schleicher and Schüll No. 589, 3, or 1505.

Procedure

(*a*) *Extraction*:

1. Weigh out between 0.25 g and 5.0 g of sample into 25 mm × 150 mm assay tubes containing not more than 40 ng B_{12}.
2. Add 20 ml of acetate buffer solution and 2 drops of sodium cyanide solution.
3. Cover the tubes with aluminium foil caps.
4. Immerse in a boiling water bath for 30 minutes.
5. Cool and dilute the extracts to 50 ml with water.
6. Filter through Whatman No. 42 paper (or No. 541 paper if samples do not contain fat).
7. If necessary the filtrate can be stored in a refrigerator overnight.

(*b*) *Preparation of inoculum*:

1. Subculture the organism overnight from the *L. leichmanii* culture and transfer to a tube containing 5 ml of sterile micro inoculum broth.
2. Incubate overnight at 37°C.
3. Mix and transfer 1 drop aseptically to a tube containing 5 ml of single strength vitamin B_{12} medium fortified with vitamin B_{12} (Note 2).
4. Incubate for 6 hours at 37°C.
5. Prepare the inoculum by transferring 1 drop with a sterile pipette to an assay tube containing 10 ml of single strength assay media.

(*c*) *Preparation of assay solutions*:

1. From the stock solution A, dilute 2 ml + 2 drops of sodium cyanide → 1000 ml with water to give standard B(20 ng/ml).
2. From standard B dilute 4 ml to 100 ml to give standard C (0.80 ng/ml).
3. Pipette 0, 0.05, 0.1, 0.15, 0.2, and 0.25 ml aliquots of standard solution C in duplicate into assay tubes.
4. Dilute sample extracts, if necessary, to give a concentration of approximately 0.80 ng/ml.
5. Pipette 0.05, 0.1, and 0.2 ml aliquots of each extract in duplicate into assay tubes.
6. Using a suitable dispenser, add 10 ml of single strength media to each assay tube.
7. Add 10 ml of media into a further 5 assay tubes to act as blanks or to use in the preparation of inoculum.
8. Cap all the tubes.
9. Autoclave the whole assay for 5 minutes at 121°C (15 lb/in²) (Note 3).
10. Remove from autoclave and cool to below 30°C in a water bath.

(*d*) *Inoculation and incubation of tubes*:

1. Loosen all caps, but leave tubes covered.

2. Fill a sterile 10 ml pipette with the inoculum.
3. Remove cap from first assay tube, then add 1 drop only of inoculum and rapidly replace the cap.
4. Repeat the process for each tube in the assay except the 5 blanks (uninoculated blanks).
5. Place the assay in the incubator for 22 hours at 37°C.

(e) *Measurement of growth*:
1. Mix contents of each set of tubes using a Vortex mixer, just before reading. Read the assay on an EEL nephelometer.
2. Place an inoculated blank tube in nephelometer and adjust the reading to zero.
3. Place an assay tube containing the top level of standard in the instrument and set the readings to approximately 95–100% full scale deflection.
4. Repeat the previous two procedures until the instrument is stable at 0 and a point between 95–100% full scale deflection.
5. Read the remaining assay tubes and record the galvonometer readings.

Calculation
1. Construct a standard curve by plotting nephelometer readings against pg vitamin B_{12} per tube for standards.
2. From the graph obtain the amount of vitamin B_{12} (pg) in each pair of assay tubes x_1, x_2, x_3.
   Let: Aliquots (ml) taken for the assay = v_1, v_2, v_3, respectively.
   Then: Concentration of vitamin B_{12} in sample extract $(x_1/v_1 + x_2/v_2 + x_3/v_3)/3 = a$ pg/ml.

Let: Weight (g) of sample $= W$
      Volume (ml) of extract $= V$
Then: Vitamin B_{12} content (μg/100 g) $= (a \times V \times 10^2)/(W \times 10^6)$
                                             $= (a \times V)/(W \times 10^4)$

Notes
1. All operations must be carried out away from direct sunlight.
2. 5 ml aliquots of single strength B_{12} media containing 0.04 ng of vitamin B_{12} per ml are dispersed into tubes and autoclaved and stored as for micro inoculum broth.

Reference
Bell, J. G. (1974). Microbiological assay of vitamins. *Laboratory Practice*, May.

8.11 Ascorbic Acid (Visual Titration Method)

Application

The method is generally applicable to foodstuffs that give colourless or faintly coloured extracts.

Principle

Ascorbic acid reduces 2,6-dichlorophenol-indophenol dye rapidly and quantitatively in acid solution. Interference from sulphite, sulphides, and thiols is removed by condensing these substances with formaldehyde at pH 0.6 before titration of the ascorbic acid. Reductones do not condense with formaldehyde under these conditions and are estimated separately after condensing ascorbic acid and other interfering substances with formaldehyde, at pH 3.5.

Reagents

Metaphosphoric acid (M.P.A.). (i) 20% solution: Dissolve 200 g of metaphosphoric acid sticks in 900 ml of water. Dilute to 1 litre and filter. Store at 0°C and prepare fresh weekly; (ii) 5% solution: Prepare a 1:3 dilution of 20% MPA with water.

2,6-Dichlorophenol-indophenol, 0.08% solution. Dissolve 0.8 g of dye in hot boiled water, filter, and dilute to 1 litre. Store in a dark bottle at 0°C.

Acetate buffer (pH 4.0). Dissolve 125 g of sodium acetate trihydrate in water and dilute to 250 ml. Add 250 ml of glacial acetic acid.

Sulphuric acid, 50% v/v.

Formaldehyde, 37–41%.

Ascorbic acid standard solution. Dissolve 0.05 g of ascorbic acid in 5% MPA solution and dilute to 200 ml with the same solvent. Prepare this solution fresh daily and use immediately to standardise the dye.

Apparatus

Burette, graduated in 0.02 ml.

Macerator; Ultra Turrax is a suitable apparatus.

Procedure

(a) *Standardisation of the dye*:
1. Dilute the dye to the required strength before standardising, that is 0.08, 0.04, or 0.02% (Note 1).
2. Pipette a 5 ml aliquot of ascorbic acid solution into a 25 ml test tube.
3. Fill a burette graduated in 0.02 ml, with the dye solution.
4. Titrate rapidly until a pink colour lasts for 5–10 seconds.
5. Titrate a 5 ml aliquot metaphosphoric acid solution, to give a value for the dye blank.

6. Subtract the dye blank titre from the standardisation titre.
7. Calculate the dye factor F (mg ascorbic acid equivalent to 1 ml of dye solution).

(b) *Extraction of sample* (Note 2):
  1. Macerate a weighed sample (Note 3) having known moisture content with 350 ml of 5% MPA for 3 minutes.
  2. Fill a 200 ml centrifuge cup with the solution.
  3. Centrifuge for 15 minutes at 2500 r.p.m.
  4. Pipette a 25 ml aliquot into a 50 ml beaker.

(c) *Titration of ascorbic acid* (Note 4):
  1. Add 1 ml of 50% sulphuric acid to the beaker, to reduce the pH to 0.6.
  2. Add 2.9 ml of formaldehyde.
  3. Pipette two 10 ml aliquots into a 25 ml test tube.
  4. *Exactly* 8 *minutes* after adding the formaldehyde, titrate with the dye.
  5. Repeat stages (c)1–(c)4 on 25 ml of 5% MPA to determine the dye blank.
  6. (If reductones present). Pipette 25 ml of sample extract from b.4 into a 50 ml beaker.
  7. Adjust to pH 3.5 with a known volume of acetate buffer.
  8. Pipette two 10 ml aliquots into a 25 ml test tube.
  9. Add 10 ml of formaldehyde.
  10. Stand for 10 minutes.
  11. Titrate with 0.02% standardised dye until a pink colour persists for 5–10 seconds.
  12. Repeat stages (c)6–(c)11 on 25 ml of 5% MPA to determine the dye blank.
  13. Subtract the dye blank from the sample titration value to give the volume of dye equivalent to half the reductone content of 10 ml of buffered extract.

Calculation

(a) *If titration instructions (c)1–(c)5 are used:*

Weight (in g) of sample taken $= W$
Total volume (in ml) of extract (volume of extractant (350 ml) plus moisture content of sample $= S$
Volume (in ml) of extract (25 ml) sulphuric acid and formaldehyde at pH 0.6 $= 28.9$
Volume (in ml) of dye corresponding to the ascorbic acid content of 10 ml aliquot $= T_1$
Dye factor for dye used $= F$

$$\text{Ascorbic acid (mg/100 g)} = \frac{T_1 \times F \times 28.9 \times S \times 100}{10 \times 25 \times W}$$

(b) If titration instructions c.6–c.12 are used in conjunction with c.1 to c.5.:

Volume (in ml) of dye corresponding to
half the reductone content of 10 ml aliquot $= T_2$
Volume (in ml) of extract (25 ml and
acetate buffer at pH 3.5 $= V_2$
Dye factor of 0.02% dye $= F^1$
Other symbols as in section (a) calculation.

$$\text{Ascorbic acid (mg/100 g)} = \left(\frac{T_1 \times 28.9 \times F}{10} - \frac{2T_2 \times V_2 \times F^1}{10}\right)\left(\frac{S \times 100}{25 \times W}\right)$$

Notes
1. Use 0.08% for samples containing more than 25 mg ascorbic acid/100 g.
   Use 0.04% for samples containing 10–25 mg ascorbic acid/100 g.
   Use 0.02% for samples containing less than 10 mg ascorbic acid/100 g if 50 g samples are extracted in 350 ml of MPA.
2. Sample extraction and titration must be completed as rapidly as possible.
3. Weigh 50 g. Dehydrated samples must be re-hydrated as follows: weigh 10 g, soak in 100 ml of 5% MPA for 20 minutes, then add a further 250 ml of 5% MPA and extract.
4. Samples that have been stored for long periods or extensively heated in processing may contain reductones. These samples should be titrated as in c.6–c.12.

Reference
Journal Soc. Chem. Ind. (1943). **62**, 223.

8.12 Ascorbic Acid (Electrometric Method)

Application
Fruit juices, fruit concentrates, jams, and vegetables producing highly coloured extracts.

Principle
Ascorbic acid polarises the cathode in a titration cell containing two platinum foil electrodes. No current flows through the cell until excess 2,6-dichlorophenol-indophenol dye is present to depolarise the cathode.

Reagents
Extracting solution. Dissolve 30 g of crushed metaphosphoric acid in 40 ml

of glacial acetic acid and 300 ml of water and add 1.9 g of EDTA. Shake this mixture at room temperature until dissolved and dilute to 1 litre with water. Store in a brown bottle at 5°C. Prepare fresh weekly.

2,6-Dichlorophenol-indophenol solution (DCIP). Dissolve 50 mg of the sodium salt in approximately 150 ml of hot water containing 42 mg of sodium bicarbonate. Dilute this solution to 200 ml and store in a brown bottle at 5°C. Standardise the dye prior to use against standard ascorbic acid solution (Note 1).

Standard ascorbic acid solution. Weigh to the nearest 0.1 mg approximately 20 mg of ascorbic acid, dissolve in extracting solution, and dilute to 100 ml.

Apparatus

Potentiometric titrator. Metrohm Potentiograph E436 or other suitable instrument with constant voltage supply.

Platinum electrodes; two platinum foil electrodes each 1 cm square.

Titration vessel, 150 ml capacity.

Burette, 10 ml graduated 0.02 ml.

Filter paper, Whatman No. 541 or Schleicher and Schüll No. 1505.

Procedure

(a) *Standardisation of the dye*:

1. Pipette a 5 ml aliquot of standard ascorbic acid solution into the titration vessel containing 40 ml of the extracting solution.
2. Connect the potentiometric apparatus and apply a constant voltage of 50 mV across the electrodes (follow instructions for particular apparatus used).
3. Titrate with DCIP from the burette at a steady rate of 2 ml/min.
4. Continually monitor the current flow through the cell.
5. Continue titrating until 1–2 ml after sharp increase in current flow (Fig. 24).
6. Extrapolate the current rise to the baseline to obtain the equivalence point (C).
7. Carry out a blank determination taking 5 ml of extracting solution and following steps a.1–a.6.
8. Subtract blank titre from standard titre.
9. Calculate the dye factor F (mg ascorbic acid equivalent to 1 ml of dye solution).

(b) *Extraction of sample*:

1. Weigh to the nearest 0.1 g, 25 g of sample into a 250 ml beaker.
2. Add 120 ml of extracting solution and macerate to give a smooth suspension.

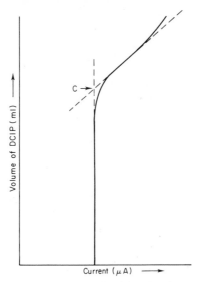

FIG. 24. Typical titration curve obtained by the electrometric procedure.

3. Filter the suspension into a 200 ml graduated flask and wash thoroughly
   with a further 60 ml of extracting solution (Note 2).
4. Make up to volume with extracting solution.

(c) *Titration of sample*:
1. From the sample extract, pipette an aliquot (up to 50 ml), containing 1 mg
   of ascorbic acid. Adjust volume to approximately 50 ml with extracting
   solution, if necessary.
2. Titrate as described under a.1–a.6 and determine the equivalence point.

Calculation

Let: Weight (g) of sample $= W$
       Volume (ml) of extract for titration $= V$
       Titre (ml) of DCIP for extract $= a$
       Titre (ml) of DCIP for blank $= b$
       Dye factor (mg ascorbic acid per ml DCIP) $= F$

Then: Ascorbic acid (mg/100 g) $= \dfrac{(a-b) \times F \times 20{,}000}{V \times W}$

Notes
1. Decomposition products that make the end point indistinct develop in
   time in the titrant. Add 15 ml of standard ascorbic acid solution to 15 ml
   dye reagent. If the solution does not become practically colourless, discard
   and prepare a new stock of solution.

2. If difficulties are encountered filtering the suspension, centrifuge for 10 minutes at 2000 r.p.m. and decant off the extract into the 200 ml graduated flask, re-suspend the solid material in a further 60 ml extracting solution, and centrifuge again.

8.13 Ascorbic and Dehydroascorbic Acid (Fluorimetric Method)

Application

The method is generally applicable to foodstuffs.

Principle

Ascorbic acid is converted into the dehydro-form by treatment with norit. The α-diketo group of this compound reacts with *o*-phenylenediamine and the quinoxaline thus formed is determined fluorimetrically. Dehydroascorbic acid can be determined directly.

Reagents

o-Phenylenediamine solution. Dissolve 11.9 mg of *o*-phenylenediamine in 25 ml of 0.01 N HCl and make up to 100 ml with water. Prepare fresh immediately before use.

Sodium acetate solution. Dissolve 500 g of sodium acetate ($CH_3COONa3H_2O$) and make up to 1 litre with water.

Boric acid–sodium acetate solution. Dissolve 3 g of boric acid in 100 ml of sodium acetate solution. Prepare this reagent fresh before use.

Extraction solution. Dissolve 60 g of crushed metaphorphoric acid sticks in 80 ml of glacial acetic acid (sp. gr. 1.04) and 300 ml of water and add 1.9 g of EDTA to the solution. Shake this mixture at room temperature until dissolved and dilute to 2 litres with water. Store in a brown bottle at 5°C. This solution remains satisfactory for 7–10 days.

Acid washed Norit. Add 1 litre of 10% hydrochloric acid to 200 g of powdered Norit and boil. After filtering transfer the residue to 1 litre water, stir and filter, washing twice more. Dry the resulting cake overnight at about 110–120°C.

Standard ascorbic acid. Weigh accurately 50 mg of ascorbic acid, dissolve it, and make up to 50 ml with extracting solution. Keep cool and prepare fresh every week.

Diluted standard ascorbic acid. Dilute 10 ml of the standard ascorbic acid to 250 ml with extraction solution. Prepare fresh every day.

Apparatus

Centrifuge, 3000 r.p.m.
Automatic syringes, up to 5 ml.
Whirly mixer, for whirling test tubes.

Fluorimeter, suitable for excitation at 350 nm and emission at 540 nm. Cuvettes for fluorescence, optical pathway 1 cm.

Procedure

(*a*) *Extraction*:
1. Weigh an amount of the sample (maximum 25 g) corresponding to about 2.5–12.5 mg of ascorbic acid in a 400 ml beaker.
2. Add directly 60 ml of extracting solution and stir carefully until a smooth suspension is obtained.
3. Filter the suspension through a folded filter into a 250 ml graduated flask and wash carefully with 175 ml of extraction solution.
4. Make up to volume with extraction solution.

(*b*) *Sum of ascorbic acid and dehydro ascorbic acid*:
1. Pipette 100 ml of the following solutions: the extracted sample (step a.4); the extraction solution; the diluted standard ascorbic acid solution.
2. Add 2 g of norit to the solutions (from step b.1).
3. Shake vigorously for 20 minutes and subsequently centrifuge for 30 minutes at 3000 r.p.m.
4. Pipette 5 ml of each of the extracted sample, extraction solution and standard solution (obtained in step b.3) to 100 ml graduated flasks each containing 5 ml of boric acid–sodium acetate solution and allow to stand for 15 minutes with occasional swirling.
5. Make up to the mark with water and mix.
6. Pipette 5 ml of the following solutions after treatment according to step (b)3: the extracted sample; the extraction solution; the diluted standard ascorbic acid solution, into 100 ml graduated flasks containing 5 ml of sodium acetate solution. Immediately make up to the mark and mix.
7. Pipette 4 ml of each of the solutions (obtained in b.5 and b.6) into test tubes covered in black paper.
8. Pipette 5 ml of *o*-phenylenediamine to each by an automatic syringe and swirl each test tube on a vortex mixer and allow to stand in the dark for 35–40 minutes. Transfer samples of the resulting solutions for fluorimetric analysis.
9. Designate the tubes as follows:
   from step b.5
   the extracted sample = A
   the extraction solution = B
   from step b.6
   the extracted sample = C
   the extraction solution = D
   the diluted standard ascorbic acid solution = E

(c) *Dehydroascorbic acid*:

1. Pipette 5 ml of the following solutions: the extracted sample a.4; the extraction solution, into 100 ml graduated flasks each containing 5 ml of boric acid–sodium acetate solution.
2. Allow to stand for 15 minutes with occasional stirring.
3. Make up to the mark and mix.
4. Pipette 5 ml of the solutions mentioned in c.1 into 100 ml graduated flasks containing 5 ml sodium acetate solution, immediately make up to the mark and mix.
5. Follow the steps b.7 and b.8 for the solutions obtained in steps c.3 and c.4.
6. Designate the tubes as follows:
   from step c.3
   　　the extracted sample = F
   　　the extraction solution = G
   from step c.4
   　　the extracted sample = H
   　　the extraction solution = J

(d) *Measurement*:

1. Set the fluorimeter according to the instructions at excitation wavelength 350 nm and at emission wavelength 430 nm.
2. Fill a cuvette with extraction solution and use as zero point.
3. Transfer the standard in acetate solution (E, step b.9) to a 1 cm cuvette and set the apparatus at 80% extinction (Note 1).
4. Transfer the solutions to a 1 cm cuvette and measure the extinction.

Calculation

Let:　Weight (g) of sample　= W

　　　Ascorbic acid + dehydroascorbic acid

　　　Readings for the H_3BO_3-treated solutions
　　　the extracted sample　= A
　　　the extraction solution　= B

　　　Readings for the non-H_3BO_3 treated solutions
　　　the extracted sample　= C
　　　the extraction solution　= D

Then: The sum of ascorbic and dehydroascorbic acid (as mg ascorbic acid/100 g)

$$= \frac{(C-D)-(A-B)}{(80-D)\times W} \times 1000$$

Dehydroascorbic acid

Let: Readings for the H_3BO_3-treated solutions
       the extracted sample $= F$
       the extraction solution $= G$

       Readings for the non-H_3BO_3 treated solutions
       the extracted sample $= H$
       the extraction solution $= J$

Then: The dehydroascorbic acid content (mg/100 g)

$$= \frac{(H-J)-(F-G)}{(80-D) \times W} \times 990$$

Note

The instrument settings were adjusted to give a transmission reading of 80%
for the standard AA solution, except when an extract solution was liable to
provide a value of 'over' 100% transmission, in which case the standard AA
value was reduced at the instrument to bring the extract solution 'on scale'.

Note added in proof
8.5 Niacin (Automated High Performance Liquid Chromatographic Method)

Notes
1. Check each batch of enzymes through the method. If these reagents give a
   blank value correct the heights of sample and standard peaks.
2. Cyanogen bromide is very poisonous, and should only be handled in a fume
   cupboard. As a further precaution put the flask containing the solution into
   a plastic bag. The waste stream should be diluted with a large excess of water.
3. This pre-treatment procedure using enzymes is the same used in the
   determination of vitamins B and B_2d so that in an automated system the
   three vitamins can be dealt with in one run.

SECTION 9

Calculation of Calorific Value

Application
The method is applicable to all foods.

Principle
The body requires energy to maintain the normal processes of life and to meet the demands of activity and growth. The unit of energy conventionally used by nutritionists is the kilocalorie (kcal). Energy intake is defined as the sum of the metabolizable energy provided by the available carbohydrate, fat, protein, and alcohol in the food ingested. Available carbohydrate is defined as the sum of the glucose, fructose, sucrose, maltose, lactose, dextrins, and starches in the diet. In calculating the energy value of a diet the contributions, if any, from other carbohydrates, e.g. cellulose, and from organic acids are ignored.

Factors
The choice of energy conversion factors is a subject of research and some controversy. In general the factors given below are used though minor differences between different countries exist.

TABLE 7

Component	Conversion factor (kcal/g)	Conversion factor (kilojoules/g) (Note 1)
Fat	9.0	37
Protein	4.0	17
Available carbohydrate (expressed as monosaccharide)	3.75	16
Starch	4.1	—
Saccharose	3.9	—
Glucose, fructose	3.75	16
Alcohol	7.0	29

The U.K. recommends the conversion factor 3.75 kcal/g for available carbohydrate (expressed as monosaccharide). Other countries have separate values for starch saccharides and monosaccharides. In practice the difference is insignificant.

Calculation

Let: Protein content (%) $= P$
       Fat content (%) $= F$
       Available carbohydrate (%) $= C$
Then: Caloric value (kcal per 100 gram) is the sum of:
       $P \times 4.0$ (protein   calories) $+ F \times 9.0$ (fat   calories) $+ C \times 3.75$
       (carbohydrate calories)
       or caloric value (kjoules per 100 g) is the sum of:
       $P \times 17$ (protein joules) $+ F \times 37$ (fat joules) $+ C \times 16$ (carbohydrate joules)

Notes
1. The precise conversion factor is: 1 kcal $= 4.184 \times 10^3$ joules or 4.184 kJ.
2. Values given in Table 7 are those proposed by the U.K. Ministry of Agriculture, Fisheries and Food, Food Standards Committee (October 1976).

Index

Acid detergent fibre, analytical method, 153, 154
Agars, 12
Air oven method, moisture and total solids, 107, 108
Alcohols, derived lipids, 17
Ale, composition table, 85
Almonds, composition table, 88
Amino acids
  composition, automated chromatographic method, 121–128
  determination, 47
  essential, 7
  limiting, 7
  structure, 6
  sulphur in, 27
Ammonia, colorimetric method, 175–178
Anaemia, 27
Apple pie, composition table, 90
Apples, composition table, 84
Apricots, composition table, 84
Arachidonic acid, essential fatty acids, 18
Ascorbic acid
  chemistry and biological role, 42
  deficiency, 42
  determination, 55
  dietary sources, 42
  electrometric method, 232–235
  fluorimetric method, 235–238
  food contribution, UK 75, 76
  recommended daily intake, 68–70
    Australia, 66
    Canada, 64

Netherlands, 58
South Africa, 65
Sweden, 63
UK, 60
USA, 61
West Germany, 62
  stability, 42
  from vegetables, 78
  visual titration method, 230, 232
  dehydro See Dehydroascorbic acid
L–Ascorbic acid, structure, 42
Ash, analytical methods, 166, 167
Asparagus, composition table, 86
Atomic absorption spectrophotometry
  method for metals, 168–172
  minerals, 52
Australia
  calcium requirements, 71–72
  protein requirements, 68
  recommended daily intakes, 66
    protein, 67–68
    vitamin C, 69–70

Bacon, composition table, 82
Bananas, composition table, 84
Beef, composition table, 82
Beetroot, composition table, 87
Benzoic acid, p–amino chemistry and biological role, 41
Beriberi, 74
Beverages, contribution to nutrient intakes in UK, 75, 76
Bicarbonates in body fluids, 23

Biological value, 7–8, 47
Biotin
  dietary sources, 41
  stability, 41
  structure, 41
  sulphur and, 24, 27
Biscuits, composition table, 89
Blackberries, composition table, 84
Black currants, composition table, 84
Blancmange, composition table, 90
Blood, clotting, prothrombin in, 34
Body fluids, control, 23
Body water balance, 20
Bones, mineral requirements, 26
Brazil nuts, composition table, 88
Bread
  composition table, 89
  contribution to nutrient intakes in UK,
    75, 76
Brussels sprouts, composition table, 87
Butter, composition table, 88
Buttermilk, composition table, 85

Cabbage, composition table, 87
Cake, composition table, 89
Calcium
  atomic absorption method, 168
  biological function, 24
  deficiency, 28
  dietary source, 24
  food contribution, UK 75, 76
  recommended daily intakes, 70–74
    Australia, 66
    Canada, 64
    Netherlands, 58
    South Africa, 65
    Sweden, 63
    UK, 60
    USA, 61
    West Germany, 62
  role, 22, 26
  source, 26
  from vegetables, 78
  vitamin D and, 31
Calories, availability, 19
Calorific value
  calculation, 19, 239, 240
  definition, 18, 19
  determination, 49
Canada

calcium requirements, 73
  protein requirements, 68
  recommended daily intakes, 64
    protein, 67
Capacitance, in moisture content
  determination, 50
Carageenans, 12
Carbohydrates
  analysis, 47, 48
  analytical methods, 130–154
  availability, 9–10
  dietary requirements, 9
  energy values, 19
  food contribution, UK, 75, 76
  recommended daily intake, Nether-
    lands, 60
  source, 9
  structure, 10–11
  thiamin and, 35
  total available, automated Clegg
    Anthrone method, 131–134
    manual Clegg Anthrone method,
    130, 131
  from vegetables, 78
α-Carotene, structure, 29, 30
β-Carotene
  automated high performance liquid
    chromatography, 188–191
  manual method, 183–188
  recommended daily intake, Nether-
    lands, 58
  structure, 29, 30
γ-Carotene, structure, 29, 30
Carotenoids, structures, biological
  activity, 29, 30
  stability, 31
Carrots, composition table, 87
Cashew nuts, composition table, 88
Cauliflower, composition table, 87
Celery, composition table, 87
Cellulose, 10
  availability, 10–12
  structure, 11
Cereals
  contribution to nutrient intakes in UK,
    75, 76
Cheese, composition tables, 85
Cheese sauce, composition tables, 86
Chemical score
  Mitchell and Block method, 128, 129

proteins, 47
Cherries, composition table, 84
Chestnuts, composition table, 88
Chicken, composition table, 82
Chicken noodle soup, composition tables, 86
Chicory, composition table, 87
Children, calcium requirements, 26
Chloride
  daily requirements, 23
  potentiometric method, 174, 175
  rapid Volhard method, 173, 174
  recommended daily intake, West Germany, 62
  role, 22, 23
Chlorine
  biological function, 24
  dietary source, 24
Chloroform–methanol extraction, fat analysis, 48, 161–163
Chocolate, composition table, 88
Cholecalciferol
  chemistry and biological role, 31
  recommended daily intake, Australia, 66
  structure, 32
Cholesterol, 7–dehydro–
  chemistry and biological role, 31
  structure, 32
Choline, chemistry and biological role, 41
Chromium
  biological function, 24
  dietary source, 24
Clegg Anthrone method, 47, 130–134
Cobalt
  biological function, 24
  dietary source, 24
  role, 27
Cod, composition table, 83
Codliver oil, composition table, 88
Coffee, composition table, 85
Cola, composition table, 85
Complementation, proteins, 7, 8
Conductivity, in moisture content determination, 50
Consomme soup, composition tables, 86
Continuous distillation, in moisture content determination, 49
Conversion factors, 81
Cooking fat, composition table, 88

Copper
  atomic absorption method, 168
  biological function, 24
  dietary source, 24
  role, 27
Corned beef, composition table, 82
Cornflakes, composition table, 89
Cornstarch, composition table, 89
Crab, composition table, 83
Cranberries, composition table, 84
Cream, composition tables, 85
Crude fibre, 47
  analytical method, 151–153
Crude protein
  automated colorimetric method, 116–119
  Macro Kjeldahl method 113–116
Cryptoxanthin, structure, 29, 30
Cucumber, composition table, 87
Custard, baked, composition table, 90
Custard powder, composition table, 89

Dairy products, contribution to nutrient intakes in UK, 75, 76
Dates, composition table, 84
Dean and Stark distillation, moisture determination, 109, 110
Dehydroascorbic acid
  determination, 55
  fluorimetric method, 235–238
  structure, 42
Density in moisture content determination, 50
Determinate errors in sampling, 44
Dextrins, 9
  availability, 12
  structure, 10
Dielectric constants in moisture content determination, 50
Dietary fibre, 11
Disaccharides, 9
  availability, 12
  structure, 10
Drying in moisture content determination, 49

Eel, silver, composition table, 83
Eggs
  composition tables, 85

contribution to nutrient intakes in UK, 75, 76
Endives, composition table, 87
Energy values, physiological, 19
Enzymes, mineral requirements, 27
Ergocalciferol
  chemistry and biological role, 31
  structure, 32
Ergosterol
  chemistry and biological role, 31
  structure, 32
Extracellular fluids *See* Intravessel fluids

FAO/WHO, calcium requirements, 72
Fats
  analysis, 48, 49
  contribution to nutrient intakes in UK, 75, 76
  deficiency, 18
  digestion, 12
  energy values, 19
  extractable, Soxhlet method, 155, 156
  food contribution, UK, 75, 76
  in milk products, Roese–Gottlieb method, 158–161
  recommended daily intake, Netherlands, 58
  from vegetables, 78
    total, Weibul method, 156
    total, chloroform methanol method, 161
Fatty acids, 12
  calcium absorption and, 26
  composition, gas–liquid chromatography method, 163–165
  recommended daily intake, West Germany, 62
  structure, 16
Figs, composition table, 84
Fish, contribution to nutrient intakes in UK, 75, 76
Flame photometry, minerals, 52
Flour, composition table, 89
Fluorine
  biological function, 25
  dietary source, 25
  recommended daily intake, West Germany, 62
Folic acid
  dietary requirements, 40

dietary sources, 40
recommended daily intake, Australia, 66
  Canada, 64
  West Germany, 62
role, 40
stability, 40
structure, 39
——, 5, 6, 7, 8–tetrahydro–, 40
Folacin, recommended daily intake, USA, 61
Food composition tables, 79–90
Fruit, contribution to nutrient intakes in UK, 75, 76

Glucose, determination, 48
Glutamic acid, pteroyl– *See* Folic acid
Glycerides, 12
  structure, 13
Glycerol, structure, 17
Glycogen, 12
Glycolipids, structure, 15
Goitre, 27
Grapefruit, composition table, 84
Grape juice, composition table, 84
Grapes, composition table, 84
Gulonic acid, diketo–, structure, 42
Gums, 12

Haddock, composition table, 83
Ham, composition table, 82
Hemicelluloses, 11–12
Herring, composition table, 83
Honey, composition table, 88

Ice cream, composition table, 90
Imino acids, structure, 7
Indeterminate errors in sampling, 44
Infrared spectroscopy in moisture content determination, 50
Inhomogeneity in sampling, 45
meso–Inositol, chemistry and biological role, 41
Intercellular fluids, electrolyte composition, 23
Intracellular fluids, electrolyte composition, 23
Intravessel fluids, electrolyte composition, 23
Inositol, 41

Iodine
    biological function, 25
    deficiency, 27
    dietary source, 25
    recommended daily intake, Canada, 64
        USA, 61
        West Germany, 62
    role, 27
Iodised salt, 27
Iron
    absorption, 27
    atomic absorption method, 168
    biological function, 25
    deficiency, 27, 28
    dietary source, 25
    food contribution, UK, 75, 76
    recommended daily intake, Australia,
        66
        Canada, 64
        Netherlands, 58
        South Africa, 65
        Sweden, 63
        UK, 60
        USA, 61
        West Germany, 62
    role, 27
    from vegetables, 78

Jam, composition table, 88
Jelly, composition table, 88

Karl Fischer method, 50
    moisture, 110–112
Kidney, composition table, 82
Kipper, composition table, 83
Kjeldahl method, 46
    protein, 113–116

Lactation, calcium requirements, 73
Lactobacillus casei, 217
Lactobacillus letchmanii, 217
Lactobacillus plantarum, 217
Lactose
    determination, 48
    enzymic method, 144–146
Lamb, composition table, 82
Leeks, composition table, 87
Lemonade, composition table, 85
Lemon juice, composition table, 84
Lemons, composition table, 84

Lentils, composition table, 87
Lettuce, composition table, 87
Lignin, 11
Linoleic acid, essential fatty acids, 18
Lipids
    analytical methods, 155–165
    dietary requirements, 12
    structure, 13–18
Lipoprotein, fat complexes, 12
Liver, composition table, 82
Lobster, composition table, 83
Locust bean gum, 12
Luff–Schoorl method, 134–138
Luncheon meat, composition table, 82

Macaroni, composition table, 89
Mackerel, composition table, 83
Macro Kjeldahl method, 113–116
Macronutrients
    analysis, 46–51
    chemistry and biological role, 5–21
Magnesium
    atomic absorption method, 168
    biological function, 24
    dietary source, 24
    recommended daily intake, Canada, 64
        USA, 61
        West Germany, 62
    role, 22, 26, 27
    source, 24
Maltose, determination, 48
Mandarins, composition table, 84
Manganese
    atomic absorption method, 168
    biological function, 25
    dietary source, 25
    from vegetables, 78
Margarine, composition table, 88
Marmalade, composition table, 88
Meat, contribution to nutrient intakes in
    UK, 75, 76
Melons, composition table, 84
Menadione
    chemistry and biological role, 34
    structure, 34
Metals, atomic absorption, 168–171
Microbiological assay techniques, for
    vitamins, 216–229
Micronutrients
    analysis, 51–55

chemistry and biological role, 22–42
Milk
  composition table, 85
  contribution to nutrient intakes in UK,
    75, 76
Milk jelly, composition table, 90
Milk products, fat in, Roese–Gottlieb
    method, 158–161
Minerals
  analysis, 51, 52
  chemistry and biological role, 22–28
  general requirements, 22
  in body fluids, 23
  in rigid structures of body, 26
  as chemical constituents of body, 26
Mitchell and Block method, 128, 129
Moisture content
  analytical methods, 107–112
  determination, 49–51
Molybdenum
  biological function, 25
  dietary source, 25
  role, 27
Monosaccharides, 9
  structure, 10
Mushrooms, composition table, 87
Mustard, composition table, 86

1, 4–Naphthoquinone, 2–methyl– See
    Menadione
——, 2–methyl–3–difarnesyl– 34
——, 2–methyl–3–phytyl–, structure, 34
Netherlands
  calcium requirements, 71
  protein requirements, 67, 68
  recommended daily intakes, 58
    vitamin C, 69, 70
Net Protein Utilisation, 9, 46
Niacin
  dietary sources, 36, 37
  food contribution, UK, 75, 76
  high performance liquid chro-
    matographic method, 213–216
  microbiological method, 221–223
  recommended daily intake, Australia,
    66
    Canada, 64
    South Africa, 65
    Sweden, 63
    USA, 61

West Germany, 62
  from vegetables, 78
Nickel
  biological function, 25
  dietary source, 25
Nicotinamide
  dietary sources, 36
  role, 37
  structure, 37
Nicotinamide adenine dinucleotide, 37
Nicotinic acid
  dietary sources, 36
  equivalents, 37
  stability, 37
  structure, 37
Nitrogen, non-protein nitrogen de-
    termination, 120, 121
Nitrogenous compounds, analytical
    methods, 113–129
Nuclear Magnetic Resonance in moisture
    content determination, 50
Nutrients
  analysis, 43–55
  recommended intakes, 56–78
Nutritional data, interpretation, 56–78
Nuts, composition table, 88

Oatmeal, composition table, 89
Olives, composition table, 84
Onions, composition table, 87
Onion sauce, composition table, 86
Orange juice, composition table, 84
Oranges, composition table, 84
Orange squash, composition table, 85
Oysters, composition table, 83
Oxalic acid, calcium absorption and, 26

Pancakes, composition table, 90
Pantothenic acid
  dietary source, 40
  recommended daily intake, West
    Germany, 62
  stability, 40
  structure, 40
Pastry, composition table, 89
Peaches, composition table, 84
Peanut butter, composition table, 88
Pearl barley, composition table, 89
Pears, composition table, 84
Peas, composition table, 87

Pectins, 12
Pepper, composition table, 86
Phosphates, role, 22, 23
Phosphoglycerides, structure, 14
Phospholipids, 12
Phosphorus
  biological function, 27
  colorimetric methods, 178–182
  dietary source, 24
  recommended daily intake, Canada, 64
    USA, 61
    West Germany, 62
  role, 22, 26, 27
  source, 26
  from vegetables, 78
  vitamin D and, 31
Phospholipids, structure, 14
Phytic acid, calcium absorption and, 26
Picric acid method for sugars, 141–144
Pineapple, composition table, 84
Plaice, composition table, 83
Plums, composition table, 84
Polysaccharides, 9
  availability, 12
Pork, composition table, 82
Potassium
  atomic absorption method, 168
  biological function, 24
  dietary source, 24
  flame photometry, 171–173
  recommended daily intake, West
    Germany, 62
  role, 22, 23
Potatoes
  composition table, 87
  contribution to nutrient intakes in UK,
    75, 76
Prawns, composition table, 83
Pregnancy
  calcium requirements, 26, 73
  iron absorption and, 27
Preserves, contribution to nutrient
    intakes in UK, 75, 76
Protein Efficiency Ratio, 9, 47
Proteins
  analysis, 46, 47
  analytical methods, 113–129
  complementation, 7, 8
  digestion, 7
  energy values, 19

food contribution, UK, 75, 76
nitrogen determination, 113, 116
quality assessment, 46
Mitchell and Block method, for
  protein quality, 128
recommended daily intake, 67
  Australia, 66
  Canada, 64
  Netherlands, 58
  South Africa, 65
  Sweden, 63
  UK, 60
  USA, 61
  West Germany, 62
source, 5
structure, 5
from vegetables, 78
Prothrombin, vitamin K and, 34
Provitamin A, determination, 53
Pyridoxal, structure, 37
Pyridoxamine, structure, 37
Pyridoxine, structure, 37

Raisins, composition table, 84
Raspberries, composition table, 84
Recommended daily allowances, 26,
  56–78
Red currants, composition table, 84
Reducing sugars, automated picric acid
  method, 141–144
References for further reading, 91–97
Refractive index in moisture content
  determination, 50
Reproduction, vitamin E and, 33
Resistivity in moisture content determination, 50
Respiration, iron and, 27
Retinol
  recommended daily see Vitamin A
  structure, 29, 30
——, dehydro–
  structure, 29, 30
Riboflavin
  food contribution, UK, 75, 76
  high performance liquid chro-
    matographic method, 210–213
  manual method, 205–207
  microbiological method, 218–221
  recommended daily intake, Australia,
    66

Canada, 64
Netherlands, 58
South Africa, 65
Sweden, 63
UK, 60
USA, 61
West Germany, 62
role, 36
sources, 36
stability, 36
structure, 36
from vegetables, 78
Rice, composition table, 89
Rice pudding, composition table, 90
Rickets, 26
Roese–Gottlieb method, 48, 158–161
Runner beans, composition table, 86

Saccharomyces carlsbergensis, 217
Salad cream, composition table, 86
Salmon, composition table, 83
Salt *See* Sodium chloride
Sample preparation, 105, 106
    composition changes and, 45
Sampling errors and procedures, 44–46
Sardines, composition tables, 83
Sauce, composition table, 86
Sausage, composition table, 82
Selenium
    biological function, 25
    dietary source, 25
Sherry, composition table, 85
Shortbread, composition table, 89
Silicon
    biological function, 25
    dietary source, 25
Sodium
    atomic absorption method, 168
    biological function, 24
    daily requirements, 23
    dietary source, 24
    flame photometry, 171–173
    recommended daily intake, West
        Germany, 62
    role, 22, 23
Sole, composition table, 83
South Africa
    calcium requirements, 71, 73
    protein requirements, 68
    recommended daily intakes, 65

vitamin C requirements, 69
Soxhlet method for fat, 48, 155, 156
Soyaflour, composition table, 89
Spaghetti, composition table, 89
Sphingolipids, structure, 15
Spinach, composition table, 87
Spirits, composition table, 85
Starch, 9
    availability, 12
    determination, 48
    enzymic method, 139–141
    Luff–Schoorl method, 134–138
    structure, 11
Starvation, iron absorption and, 25
Sterol, 12
    structure, 17
Strawberries, composition table, 84
Sucrose
    automated   picric   acid   method,
        141–144
    determination, 48
Sugar
    composition, high performance liquid
        chromatography method, 146–151
    composition table, 88
    contribution to nutrient intakes in UK,
        75, 76
    total low molecular weight, Luff–
        Schoorl method, 134–138
Sulphur
    biological function, 24
    dietary source, 24
    role, 27
Sweat, salt loss in, 23
Sweden
    recommended daily intakes, 63
    protein, 67, 68
Swedes, composition table, 87
Sweet corn, composition table, 87
Sweet potatoes, composition table, 87
Syrup, composition table, 88

Tangerines, composition table, 84
Tapioca, composition table, 89
Tea, composition table, 85
Teeth, mineral requirements, 26
Thiamin
    chemistry and biological role, 35, 36
    determination, 54
    dietary sources, 35

food contribution, UK, 75, 76
high performance liquid chromatographic method, 208–210
manual method, 201–205
recommended daily intake, Australia, 66
  Canada, 64
  Netherlands, 58
  South Africa, 65
  Sweden, 63
  UK, 60
  USA, 61
  West Germany, 62
stability, 35
structure, 35
sulphur and, 27
from vegetables, 78
Thyroid hormones, iodine and, 27
Tin
  biological function, 25
  dietary source, 25
Tocols
  chemistry and biological role, 32–34
  structure, 33
Tocopherols, chemistry and biological role, 32–34
α-Tocopherol,
  activity, 33
  recommended daily intake, West Germany, 62
β-Tocopherol, activity, 33
δ-Tocopherol, activity, 33
γ-Tocopherol, activity, 33
Tocotrienols
  chemistry and biological role, 32–34
  structure, 33
α-Tocotrienol, activity, 33
Toffee, composition table, 88
Tomatoes, composition table, 87
Total fat
  chloroform–methanol extraction, 161–163
  Weibul method, 156–158
Total nitrogen
  automated colorimetric method, 116–119
  Macro Kjeldahl method, 113–116
Total solids, analytical methods, 107–109
Treacle, composition table, 88
Treacle tart, composition table, 90

Trifle, composition table, 90
Triglycerides, 12
Turbot, composition table, 83
Turkey, composition table, 82
Turnips, composition table, 87
UK
  calcium requirements, 73, 74
  food contributions to total nutrient intake, 75–76
  protein requirements, 68
  recommended daily intakes, 59
    protein, 58
Urine, salt loss in, 23
USA
  calcium requirements, 72, 73
  protein requirements, 67–68
  recommended daily intakes, 60
    vitamin C requirments, 69, 70
Vaccuum oven method, moisture and total solids, 108, 109
Vanadium
  biological function, 25
  dietary source, 25
Vapour pressure in moisture content determination, 50
Veal, composition table, 82
Vegetables
  contribution to nutrient intakes in UK, 75, 76
  role in provision of nutrients, 78
Vinegar, composition table, 86
Vitamins (See also specific vitamins)
  analysis, 52–55
  chemistry and biological roles, 26–42
  microbiological assay techniques, 216–229
Vitamin A (See also Retinol)
  automated high performance liquid chromatographic method, 188–191
  chemistry and biological role, 29
  deficiency, 29
  determination, 53
  dietary fats and, 18
  dietary sources, 29, 31
  digestion, 29
  food contribution, UK, 75, 76
  manual method, 183–188
  recommended daily intake
    Australia, 66
    Canada, 64

Netherlands, 58
South Africa, 65
Sweden, 63
UK, 60
USA, 61
West Germany, 62
stability, 31
from vegetables, 78
Vitamin A₁ *See* Retinol
Vitamin A₂ *See* Retinol, dehydro–
Vitamin B, determination, 54, 55
other B-group vitamins, 41
Vitamin B₁ *See* Thiamin
Vitamin B₂ *See* Riboflavin
Vitamin B₆ (*See also* Pyridoxal;
Pyridoxamine; Pyridoxine)
chemistry and biological role, 37, 38
microbiological method, 224–227
recommended daily intake, Canada, 64
USA, 61
West Germany, 62
role, 38
sources, 38
stability, 38
from vegetables, 78
Vitamin B₁₂
microbiological method, 227–229
recommended daily intake, Australia,
66
Canada, 64
USA, 61
West Germany, 62
role, 39
source, 39
stability, 39
structure, 38
sulphur and, 25
from vegetables, 78
Vitamin C *See* Ascorbic acid
Vitamin D (*See also* Cholecalciferol;
Ergosterol)
calcium absorption and, 26
chemistry and biological role, 31, 32
deficiency, 32
determination, 53
dietary fats and, 18
digestion, 31
food contribution, UK, 75, 76
gas chromatographic method, 191–197
recommended daily intake

Australia, 66
Canada, 65
Sweden, 63
UK, 60
USA, 61
West Germany, 62
sources, 31, 32
stability, 32
Vitamin D₂ *See* Ergocalciferol
Vitamin D₃ *See* Cholecalciferol
Vitamin E (*See also* Tocopherols)
chemistry and biological role, 32–34
determination, 54
dietary fats and, 18
dietary sources, 34
recommended daily intake, Canada, 64
USA, 61
West Germany, 62
role, 33
stability, 34
thin layer chromatographic method,
197–200
Vitamin K (*See also* Menadione)
chemistry and biological role, 34, 35
dietary sources, 34
stability, 35
Vitamin K₁ *See* 1, 4–Naphthoquinone,
2–methyl– 3–phytyl–
Vitamin K₂ *See* 1, 4–Naphthoquinone,
2–methyl–3–difarnesyl–
Vitamin K₃ *See* Menadione
Volhard method, 173, 174

Walnuts, composition table, 88
Water
chemical reactivity, moisture content
determination and, 50
dietary requirements, 20
electrolytes in, 23
function, 20
properties, 20
recommended daily intake, West
Germany, 62
Waxes, 12
structure, 13
Weibul method for total fat, 48, 156–158
West Germany
protein requirements, 67
recommended daily intakes, 62
vitamin C requirements, 70

Wet digestion method for element
    analysis, 167
Wheat germ, composition table, 89
Wine, composition table, 85

Yoghurt, composition tables, 85

Zinc
    atomic absorption method, 168
    biological function, 25
    dietary source, 25
    recommended daily intake, Canada, 64
        USA, 61
    role, 28